CARPENTRY FOR THE HOMEOWNER

CARPENTRY FOR THE HOMEOWNER

Peter Marfleet

B.T. Batsford Ltd · London

ISBN 0 7134 4792 3

Typeset by Tek-Art Ltd, Kent
and printed by
Anchor Brendon Ltd
Tiptree, Essex
for the publishers B.T. Batsford Ltd
4 Fitzhardinge Street, London, W1H 0AH

CONTENTS

1 HAND TOOLS

The array of hand tools now available to the home carpenter is formidable, and there is a great temptation to go to the nearest handicraft centre and spend a small fortune. Beautiful tools and equipment can give a great deal of pleasure, both to look at and use, but it is quite likely that some of the tools chosen may never be used. Suppliers such as Black and Decker, Bosch, Leichtung, Marples, Shopsmith, Stanley and Wolf produce a wide variety of hand and powered tools, and in some cases a range of different qualities. It is essential to examine items carefully before purchase to ensure that they comply with your needs.

It would be impractical to detail all the tools available in a single chapter, but guidance is given on the various groupings, and those tools that are most likely to be needed for jobs around the home.

Cutting Tools

The tools falling into this category are:

axe	knife	saw
chisel	plane	scraper
file	router	spokeshave

The axe (fig. 1) is not normally considered to be part of the standard tool chest, but it is certainly a very useful tool, and not just for cutting firewood!

Small pegs, wedges and plugs are most easily fashioned from off-cuts by using the axe for the initial shaping. Indeed, for a lot of rough work the axe is ideal, and it can also be used to provide an interesting texture for a 'rustic' look.

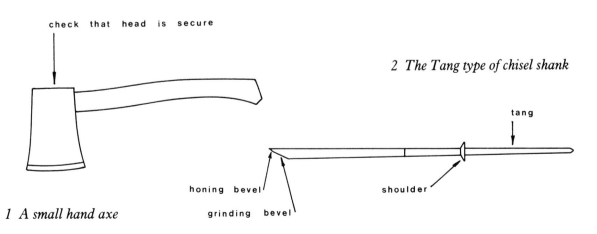

check that head is secure

2 The Tang type of chisel shank

tang

honing bevel

grinding bevel

shoulder

1 A small hand axe

Care must be taken to ensure that the head is securely attached to the handle; there is always the tendency for the wooden shaft to dry out and shrink. This problem may be alleviated by adding a few drops of linseed oil into the end grain at the head to keep it moist.

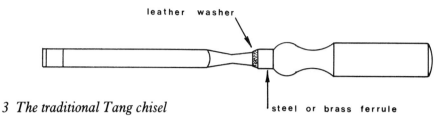

leather washer

steel or brass ferrule

3 The traditional Tang chisel

Chisels are probably the most widely used of all hand tools; they are virtually indispensible in the construction of most joints. The traditional chisel (fig. 2), broadly classified as a Tang chisel (meaning that there is a projecting shank forged on the end of the blade) has now been supplemented with the socket chisel which is more suited to heavy work.

Modern chisels generally have high-impact plastic handles moulded into a socket blade, and some manufacturers claim them to be virtually unbreakable.

The traditional Tang chisel (fig. 3) is easily distinguished, as there will be a brass or steel ferrule and leather washer between the end of the handle and the blade. This will help absorb any force when the end of the handle is hit with a mallet.

Chisels may be further classified into:

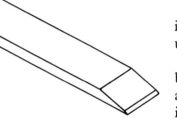

4 The firmer chisel blade

firmer	mortise
bevelled and paring	gouges

The firmer chisel (fig. 4) is really the most useful. It comes in a variety of widths, ranging from 3 to 50mm (⅛ to 2in), but its robust construction makes it ideally suited for general work.

The bevelled edge chisel (fig. 5) is lighter in construction than the firmer chisel, and blade is bevelled along its length on both edges. It is generally used for lighter work; in particular for the construction of dovetail joints.

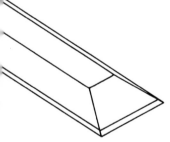

5 The bevelled edge chisel blade

7 A typical gouge blade

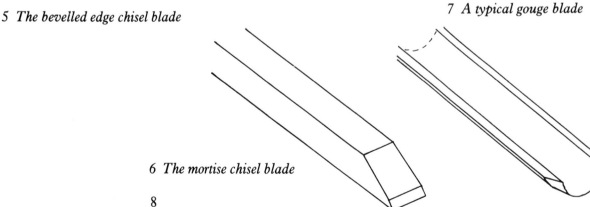

6 The mortise chisel blade

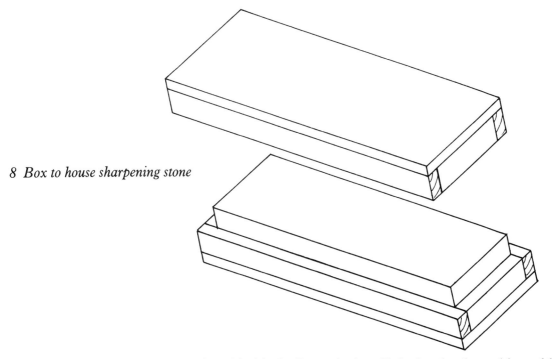

8 Box to house sharpening stone

The paring chisel is similar to the bevelled edge, but has a thinner blade and is much longer. It is ideally suited for the removal of waste wood from long grooves and is a very useful addition to the normal range of chisels.

The blade of the mortise chisel (fig. 6) is very much thicker and stronger than other chisels. This enables it to be firmly struck with a mallet, and allows it to withstand the pressure put on the blade when used to lever out the waste wood from a mortise. A ferrule is often fitted to the end of the handle to prevent it splitting under the impact of the mallet blows.

Gouges (fig. 7) are used for making grooves, flutings and cutting out hollows and depressions. They are further sub-divided into firmer, scribing, crank and spoon. Each has its own specific use, but they do not form an essential part of the home carpenter's kit. Grinding and sharpening of chisels often needs to be carried out, as chisels should always be kept sharp and well-oiled. They should be kept in a canvas wallet with seperate pouches for each chisel, so that they can be stored together without damage to the edges. A sharpening stone should be available, preferably with a medium and fine combination. This may be housed in a simply constructed box (fig. 8) that can be opened on either side to expose the grade of stone required. Grinding should be done to 25° and sharpening on the oilstone to between 35° and 40°.

Although wooden planes are still available, the modern trend is for metal planes, and there are many makes and styles available.

In buying a plane, there are several points to consider; the first being to decide on the work to be done and finding a plane suitable. It should be well balanced and feel comfortable to hold. The body should be of good quality cast iron, with the sole machined to give a flat and even surface. The blade will normally be of good quality tool steel, hardened and tempered and capable of taking and retaining a sharp edge.

9

The various adjustments will include the lever cap, lateral adjustment lever, adjusting screw and frog screw, all of which should be readily accessible and have a firm positive movement.

Metal planes can be generally classified into *jack*, *smoothing* and *block*, and an example of each is an essential item in a normal set of carpenter's tools. There are several special purpose planes available, the most useful being a multi-purpose plane that can cut rebates and grooves, a set of interchangeable blades being provided. A rebate plane and a plough plane can form a useful addition, but they are not cheap and need only be considered if a lot of joinery work is anticipated.

The spokeshave, originally designed to shape the spokes for the wheels of horse-drawn vehicles, is now made of metal, and can be useful to finish a curved edge to the required shape.

The Surform plane is available in several styles. The body of the plane is a hollow frame, and the multi-toothed blade takes the place of the sole of the traditional plane. The cutting action is similar to a file, but the honeycombe construction of the blade allows the shavings to pass through and not clog the blade. Blades are available to cut plastic, metal and masonry as well as wood, and are particularly suited for shaving down man-made boards, such as chipboard. They are relatively cheap, and are to be recommended for work that may damage the more traditional tools. Files and rasps can be useful for the preliminary shaping, in particular for awkward curved surfaces, and should be purchased as and when the need arises.

The final broad group of cutting tools are the saws, and it is essential that these are of good quality and kept sharp and oiled.

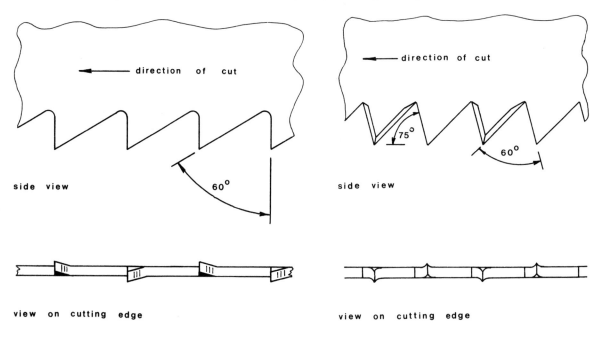

9 *Details of the rip saw blade*

10 *Details of the cross-cut blade*

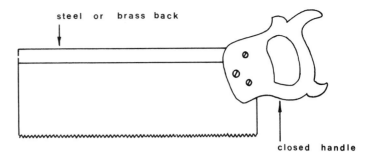

11 The tenon, or back, saw

The rip saw (fig. 9), used for cutting with the grain, is essential for reducing the width of timber as the teeth are set to cut a wider kerf than other saws, thus reducing the possibility of the saw being pinched by the wood. The cutting action is similar to a row of chisels placed directly one behind the other, each tooth removing a shaving the full width of its cutting edge.

The cross-cut saw (fig. 10) is used for cutting across the grain; the teeth are bevelled on the front of their cutting edges, giving an action similar to a series of knife-edges which sever the fibres of the wood.

The panel saw is similar to the cross-cut, but with smaller teeth and closer spacing. Its main use is for cutting thin boards or panels up to 20mm (¾in) in thickness.

The tenon or back saw (fig. 11), has a steel or brass back fitted to the blade to keep it in tension and provide weight to assist with the cutting action.

The dovetail saw is similar to the tenon saw, but with smaller teeth, and is used mainly for cutting dovetails when making cabinet drawers and other fine work.

The traditional bow saw has a deep beechwood frame and a narrow blade, and is designed for cutting curves on shaped work.

The coping saw is used for cutting smaller and sharper curves when the bow saw would be too unwieldy.

The pad saw, or keyhole saw, has a very narrow blade and can be used to cut small inside curves. It is advisable to use as short a blade as possible in order to minimize the tendency for the blade to snap or flex.

Measuring and Marking out

The try-square (fig. 12) is used to test if surfaces are square with each other, and to draw lines at right angles to a true edge or surface. The stock is made from a hard wood and faced on the inside with a strip of brass to avoid wear. The blade should be made from hardened and tempered high carbon steel and must be handled carefully to avoid damage to the edges and squareness.

The mitre square, or bevel (fig. 13), is similar to the try-square, but with the blade set at 45° to the stock.

A marking knife is used for drawing lines across the grain of the wood,

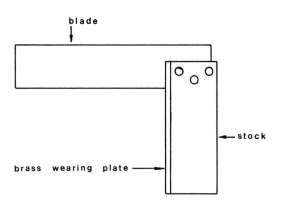

12 The try-square

and, for accurate work, is preferred to a pencil.

A marking gauge is used for drawing lines parallel to an edge or surface and for marking out correct width and thickness.

The mortise gauge has an additional spur, and is used for the setting out and marking of mortise and tenon joints.

A set of dividers, a spirit level and a straight edge are all useful additional aids in the setting out of work. A steel rule, a retractable rule and a folding wooden rule are further items that should be included.

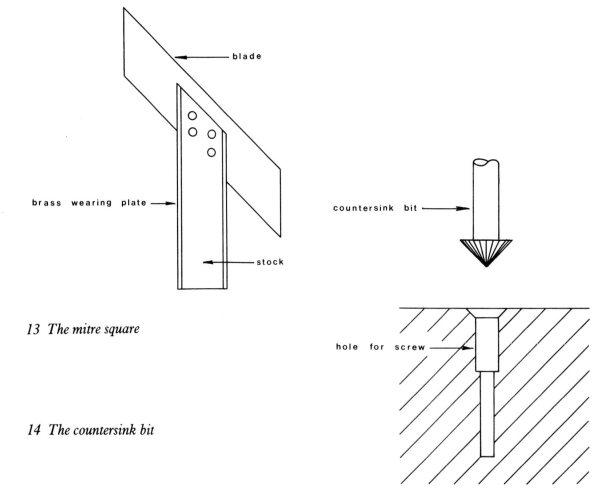

13 The mitre square

14 The countersink bit

Fixing and Holding

Various types of clamps and cramps are needed for joinery work, and are used to hold down work to the bench top when sawing, chiselling or shaping, as well as holding glued components in position whilst the adhesive sets.

The sash cramp is used for assembling large work such as window sashes, cabinets, doors and framed constructions. The G-clamp is a multi-purpose clamp, but should have protective pieces of scrap wood placed between the work and the face of the clamp to prevent the work being marked or dented.

A claw hammer is not only used for hammering nails; the claw is designed to help withdraw them. The cross-pein hammer is used for general work with the cross-pein being used to start driving in short nails and oval brads.

A wooden mallet is used for striking chisel handles when heavy cutting is needed, for example in the making of a mortise.

A selection of screwdrivers completes the fixing group, a ratchet, or American screwdriver being particularly useful as the blades can be interchanged to suit the different sizes and types of screws.

Boring

A hand, or breast drill is used for small holes – whilst the crank brace, fitted with a ratchet, is more suited to the larger holes and heavier work.

A wide variety of bits are available, each for a particular use and purpose. The basic auger, or twist bit, has a spiral point which draws the bit into the wood. The spur cutters sever the fibres of the wood and prevent splitting as the router cutters remove the waste wood from the hole.

When boring right through a piece of timber, the bit should not be allowed to break through completely, as this will cause the wood to splinter. When the point of the bit appears on the other side of the wood, the bit should be withdrawn, and the hole completed by drilling through from the other side.

A countersunk bit (fig. 14) should be used after drilling a hole that is going to take a countersunk screw, so that the maximum efficiency can be obtained from the screw.

2 POWER TOOLS

Power tools can greatly cut down on the time taken to do some of the more routine jobs, particularly if they are of a repetitive nature.

There are many types of power tools, and, although it is possible to buy attachments to enable the basic drill to perform many different tasks, there is no substitute for buying a power tool designed specifically for one purpose (fig. 15). For example, saw attachments to a drill place a completely different set of forces on to the main axle of the drill to those generated when it is used for drilling.

Circular saws

A circular saw will have a more powerful motor than the normal electric drill, and the speed of rotation will be much greater.

Some are hand held (fig. 16), others designed to be fitted into a bench table. For the home carpenter, a hand-held circular saw with a 200 to 225mm (8 to 9in) blade will cope with virtually all the work associated with jobs around the home.

Many of the hand-held saws can be mounted onto a frame so that a bench saw is obtained. Unfortunately they are not particularly sturdy, and modifications may be needed to ensure better stability.

The bench saw consists of a flat table-top through which the saw blade projects; wood is pushed over the table-top towards the blade (which can be adjusted for depth and angle) and the wood is cut accordingly. Guide

15 A hand-held belt sander

16 A portable circular saw

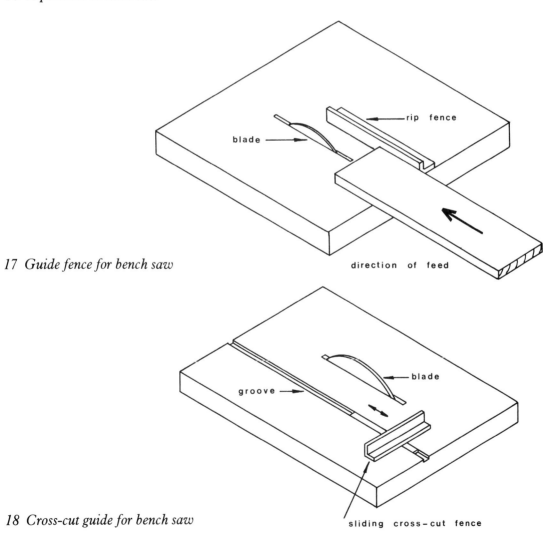

17 Guide fence for bench saw

18 Cross-cut guide for bench saw

15

fences (fig. 17) can be fitted to the table-top, the rip fence being parallel to the saw blade, the cross-cut (fig. 18) guide being able to move from the front of the table to the back by sliding in a groove that is parallel to the blade.

Great care has to be taken with all machines, and adequate guards should always be used. When using the saw bench, it is particularly bad practice to feed small pieces of wood into the blade by hand. The blade has a tendency to grab the wood out of your control. Timber should always be fed to the blade by using a push-stick, keeping the hands and fingers well away from the blade.

Mitring can be easily done, either by tilting the blade through 45°, or by setting a protractor on the cross guide to the same angle.

Firring, or tapered pieces, may be cut by making a jig that is fixed to the rip fence and set at the required angle.

Housing joints (fig. 19) are also easily cut by setting the blade to protrude the depth of the housing, and running the timber across the blade at small increments to allow the removal of the wood.

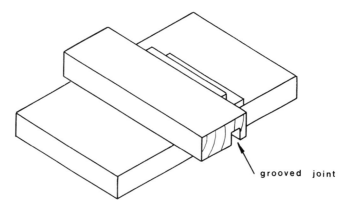

grooved joint

19 Cutting housing joints

Rebates can be cut by setting the rip fence close to the blade, running the timber through, and then turning the wood through 90° and running it through again to complete the other side of the cut.

Kerfing, that is a series of parallel cuts (fig. 20), may be carried out to allow the easy bending of a piece of timber. This will obviously greatly reduce the strength of that piece of timber, and it should therefore not be used for load bearing members.

width

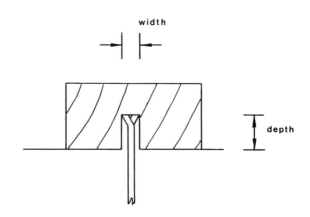

depth

20 Details of the kerf

Various types of blades are available for circular saws; rip, cross-cut or combination being the most popular. However, it is possible to purchase a tungsten carbide tipped blade, which, although cutting a slightly wider groove, will give excellent edges to the cuts. If treated carefully, the blade will last a remarkably long time and will repay the additional cost.

Drills

One of the first power tools that the home carpenter will have bought is an electric drill, although it is probable that a more powerful drill may be needed. Advances in design and efficiency make it well worthwhile considering the purchase of a heavier and stronger machine, probably with the hammer attachment.

A good selection of drill bits is advisable, whatever the drill. Each type of bit has a specific purpose, and, when correctly used, will make work a lot more efficient.

A set of twist bits for the smaller holes, a set of flat bits for the larger, and a collection of Forstner bits, bought as and when required, will serve for most jobs. Plug cutters, shell augers, Jennings bits and hole saws can all be added when a specific need occurs.

Routers

A router is a very useful addition to the home carpenter's range of power tools, as it can be used for cutting basic joints, as well as providing a professional finish to many items (fig. 21).

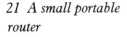

21 A small portable router

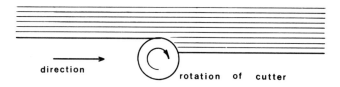

direction   rotation of cutter

22 Correct cutting action for the router

The router consists of a powerful electric motor, held vertically, with a purpose-made cutter attached to the central axis.

The cutting has to be carried out in the correct direction, and the use of fences or guides is often required to keep the cutting confined to the area concerned (fig. 22).

Some cutting tools have a rubbing guide below the cutting edge, but care has to be taken in using these to avoid burning the wood. This can also happen if the router is moved too slowly. The cutting edge will generate too much heat due to friction, and, as well as burning the wood, will blunt the cutting edge.

It is simply a matter of practice to move the router quickly enough to make a continuous cut, without putting an unnecessary strain on either the bit or the motor by trying to go too quickly.

Deep cuts should not be attempted in one pass, but rather as a series of shallow cuts. This will avoid putting too much strain on the motor (fig. 23). As there is always the risk of flying particles, the wearing of safety glasses or goggles is recommended whilst using the router.

23 Typical grooves cut using a router

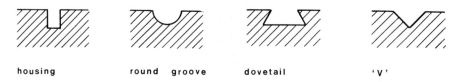

housing   round groove   dovetail   'v'

Planers

Hand-held planers use a rotary cutting block and actually produce chippings as opposed to shavings. They are especially suited to the removal of waste wood very quickly; some of the more powerful machines can take up to 3mm (⅛in) at one pass. The depth of cut can be determined by setting a pre-select gauge at the front of the machine.

Jig-saws

A jig-saw, although capable of cutting a straight line (this can be done with the aid of a temporary fence) is generally used for cutting curves and circles. With the appropriate blade, a wide range of materials can be cut. The correct blade needed will be specified in the individual manufacturer's handbook.

It is always advisable to pre-drill a hole into which the blade can fit before starting cutting operations.

3 TIMBER

The classification of timber into hardwood and softwood is often misleading, as the terms do not describe the wood's hardness or softness.

In general, wood that is classified as a hardwood comes from a tree that has broad leaves and is deciduous; that is, it sheds its leaves each year. Softwood comes from the trees with scale-like leaves, the conifers; trees such as the cedar or the pine.

The section through a log (fig. 24) shows the sapwood, i.e. the outer portion that is the living part of the tree, and the heartwood (with the inactive cells) forming the centre of the log.

Sapwood is usually lighter in colour than the heartwood; this is particularly noticeable in the English Yew. Timber is generally converted into planks etc., by two alternative methods; plain-sawn (fig. 25), or quarter-sawn (fig. 26), the latter providing better quality timber. As a method, it is generally more expensive, but it is the preferred method for timber that will be used for furniture construction and cabinet making.

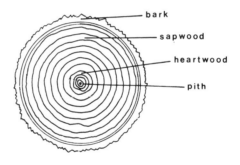

24 Section through a typical log

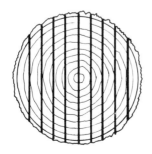

25 Plain sawn conversion

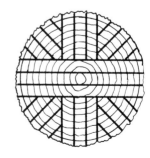

26 Quarter sawn conversion

19

Characteristics

Each type of timber has its own particular characteristics, and a summary of some of these is given below, related to the more common timbers.

Softwoods

Name of timber	Ease of working	Characteristics painting	nailing	Strength bending	stiffness
Western Red Cedar	A	A	C	C	C
Douglas Fir	C	C	B	A	A
Western Hemlock	B	B	B	B	A
Western White Pine	A	A	A	B	B
Englemans Spruce	B	B	C	C	C

Hardwoods

Name of timber	Ease of working	Characteristics painting	nailing	Strength bending	stiffness
Ash	C	C	A	A	A
Birch	C	B	A	A	A
Elm	C	C	B	A	A
Maple	C	C	B	A	A
Oak	C	C	A	A	A
Walnut	B	C	B	A	A

Ease of working is based upon hardness, texture, and character of the surface that can be obtained, with those classified as A having a soft uniform texture, which can be finished to a smooth surface.

Paint adheres better on edge grain than end grain, and in general, some preparation will be needed on the knots if the paint is to cover effectively.

Nailing quality is related to the density and splitting tendency.

Bending strength gives an indication of the relative load carrying capacity. This is explained in more detail in the chapter on design.

Stiffness refers to the ability of a member to carry the load placed on it without undue deflection or sag.

Unfortunately, due to the variable nature of timber, defects often occur, either naturally (figs 27, 28) or during the drying out and milling process (fig. 29). Most of these defects are relatively obvious, and the more common are shown in the illustrations. Others include decay due to the growth of fungus, holes from a variety of causes, pitch pockets and between the annual rings caused by pitch accumulation and discoloration.

27 Common defects in sawn timber

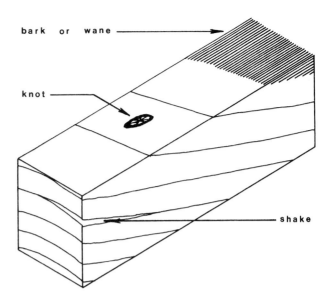

bark or wane

knot

shake

28 Common defects in log form

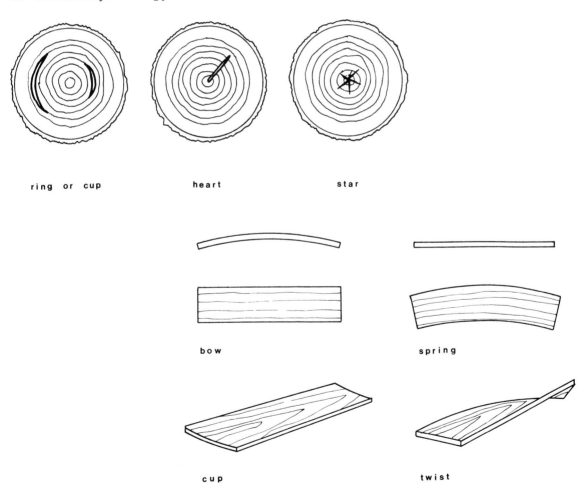

ring or cup

heart

star

bow

spring

cup

twist

29 Common defects in planks and boards

Seasoning

The method of drying out or seasoning the timber will have an effect on its workability, and some knowledge of moisture content may help towards an understanding of the problem.

In its natural state, green timber has water, both in the cell cavities, and absorbed into the cell walls. As the wood begins to dry, the cell walls remain saturated until the 'free water' in the cavities has evaporated.

For most timbers, this is achieved at around 30% moisture content. The moisture content is found from

$$\% \text{ moisture content} = \frac{\text{weight of green wood} - \text{oven dry weight}}{\text{oven dry weight}} \times 100$$

The idea is to get the timber at the percentage moisture content that it will be subjected to when installed. For example, if it is tongued and grooved panelling and will be fitted into a centrally heated house, then the timber should be kept at those conditions for several weeks before being installed. Although this is not always practical, it means that the timber will have strunk as much as it needs to, and that the finished work will be reasonably stable (fig. 30).

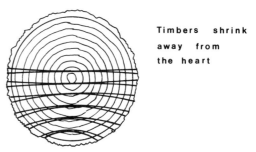

Timbers shrink away from the heart

30 Shrinkage of timber

For the serious home carpenter, it is well worth buying timber in log form, having it cut at a local sawmill, and then seasoning it oneself.

There is a theory that the stacking of unseasoned timber so that the planks lie in a north-south direction will improve the quality of the process. Whatever truth there may be in this theory, it does add to the mystique and pleasure of helping nature change one of its products into something that can last for centuries.

Another advantage of seasoning one's own timber is that it is often possible to get the trunk of an unusual tree that might have been cut down in the neighbourhood, and this may well come your way if it is known that you will be able to be put to a useful purpose instead of it just being burnt by some building contractor.

The natural method of air seasoning requires the timber to be stacked in such a way that air can circulate freely around the boards (fig. 31). The circulation will be partly controlled by the thickness of the stickers placed between the boards, and partly dependent on the location of the stack (fig. 32).

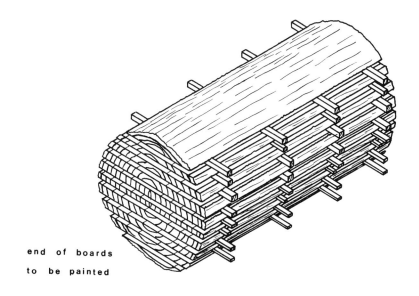

end of boards
to be painted

31 Natural seasoning of planks

note vertical airways 25 to 50 mm

32 Natural seasoning of other sections

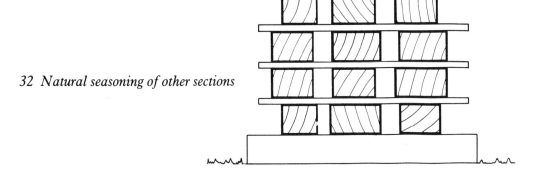

As protection against the sun and weather, a waterproof covering should be placed over the stack, and the stack should be at least 450mm (18in) above the ground, which should be kept free of weed etc. (fig. 33). The ends of each board should be sealed – petroleum grease is recommended – but a good quality aluminium primer paint will suffice. It is necessary to carry out this basic protection, for if not, the ends will dry out too quickly and in all probability will split, thus reducing the useful amount of timber available. The time taken for timber to season in the open will depend upon the type of timber, sectional size of boards, the prevailing weather, and the location of the stack. A broad rule of thumb is that it will take 12 months for every 25mm (1in) of board thickness for the timber to reach the optimum air moisture content of between 15 and 20%.

23

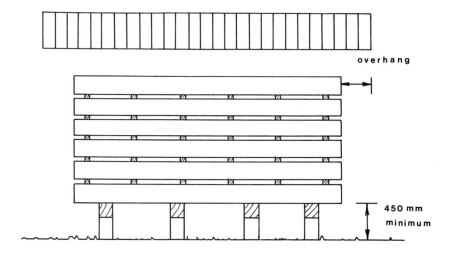

overhang

450 mm
minimum

33 *Protection during seasoning*

For structural timber, this will be adequate, but for timber being used inside for joinery, a further period of drying out will be needed to allow it to reach the same percentage moisture content as its final location. Most timber purchased from a timber yard will have been kiln dried, and, although this generally produces satisfactory results, it is good practice to examine any timber before purchasing it to ensure that it is reasonably straight and true, and will be suitable for the use intended.

Stock sizes

The normal softwood used for building construction will come from a range of six to eight basic 'pine' species, which are generally imported as sawn stock in a fairly wide range of sizes.

The distribution of sizes available will vary geographically, and it is advisable to check with your local supplier regarding his particular range. In Europe, trees are generally smaller than those of, say, Western Canada, and hence the sawn timber is not available in large sections.

Widths are rarely more than 250mm (10in) and the price of anything wider than 175mm (7in) is much higher than the smaller widths.

Most sections are available in lengths up to 5.1m (16ft) at a normal timber yard. Longer lengths may require a special order.

Comparison of stock sizes

English (measured in millimetres)

Thickness	75	100	125	150	175	200	225	250	300
				Width					
16	O	N	N	N					
19	N	N	N	N					
22	N	N	N	N					
25	N	N	N	N	N				
32	N	N	N	N	N				
38	O	N	N	N	N	O	N		
44	O	O	O	O	O	O	O	O	O
50	N	N	N	N	N	N	N	O	O
63		N	N	N	N	N	N		
75		N	N	N	N	N	N	O	O
100		O		N		N		O	O
150				O		O			O
200						O			
250								O	
300									O

N – normally available
O – special order
(other sizes within the range may be cut to suit).

American (measured in inches)

Items	Thickness		Face width	
	Nominal	Dressed	Nominal	Dressed
Boards	1	³/₄	2	1½
	1¼	1	3	2½
	1½	1¼	4	3½
			5	4½
			6	5½
			7	6½
			8	7¼
			9	8¼
			10	9¼
			11	10¼
			12	11¼
			14	13¼
			16	15¼

American (measured in inches)

Items	Thickness		Face width	
	Nominal	Dressed	Nominal	Dressed
Dimension	2	1½	2	1½
	2½	2	3	2½
	3	2½	4	3½
	3½	3	5	4½
	4	3½	6	5½
	4½	4	8	7¼
			10	9¼
			12	11¼

In contrast to softwoods, hardwoods are normally imported in random widths, and it may therefore be difficult to obtain large quantities of sawn hardwood in one width. Any width over 150mm (6in) is also likely to be more expensive.

The conversion of logs varies in different countries, although within each country the sizes are fairly standard. For example, logs in East Africa tend to be smaller than those in West Africa, so it is often helpful to consider the country of origin if looking for a particular size.

Timber from Malaysia is exported as sawn stock, the principal timbers being Keruing and Red Meranti. These will generally be 150 to 200mm (6 to 8in) with lengths averaging 3.6 to 4.2m (11 to 14ft).

From Europe, beech and oak are the main exports, with the timber sawn to some 180 to 190mm (7 to 10in) in width, and between 1.8 and 3m (6 to 10ft) in length.

In England, although timber is available in log form, much of it is cut into 25mm (1in) and thicker boards, but the sizes vary considerably with the species.

As the trade is somewhat specialized (for example one supplier may deal solely in timber for fencing and gates) it is sometimes difficult to find a source for the exact material required.

In America, elm is available from 19 to 150mm (¾ to 6in) and sometimes more, in width, and from 25 to 100mm (1 to 4in) in thickness.

Mahogany is generally available at an average width of 225mm (9in) whilst Rock Maple is available from 150mm (6in) and between 25 and 100mm (1 to 4in) in thickness.

In all cases, a visit to the local timber merchant will give precise information as to what is in stock, and what can be obtained with a special order.

Quality

Apart from its structural properties, there are three other qualities of timber worth noting.

Thermal conductivity of timber compares well with most other materials used in building construction; for example, an all-timber stud wall has a total U value 2½ to 3 times better than a brick cavity wall. Acoustically, when used in various forms in conjunction with other absorbent materials, it can provide optimum absorption throughout the audible range as well as allowing its use as a decorative finish. (Absorption in this context, means the ability to reduce reflections of sound, and timber is excellent for this purpose.)

With careful detail and treatment, timber can usually be made to conform with local fire regulations, and, on structural members, there is generally enough 'sacrifical' timber available to provide the necessary resistance. Timber chars at a predictable rate of 0.6mm per minute, (¼₀th of an inch per minute) and can therefore be designed in sizes to suit the regulations.

Selection of timber

In considering the type of timber to use, it is important to consider the following:

> *Will it be in contact with the ground?*
> *Will it be wholly enclosed in brickwork, concrete or masonry?*
> *Will it have adequate ventilation?*
> *Will it be liable to fungal or insect attack?*
> *Will it be subject to moisture attack?*

If any of the above items apply, then it will be necessary to consider using durable or very durable timber, and to apply a suitable preservative. There are three main types: tar oils, water-borne and organic solvents, each having their advantages. All three types are effective against decay and other wood destroying agents.

Whichever type is chosen, the timber must be in a suitable condition to absorb the preservative; that is, the timber should be seasoned, and at a moisture content appropriate to the preservative. Any cutting, shaping or boring should be carried out before treatment.

The finish to be applied will depend on the use and location of the timber, but, whichever finish is to be applied, the surface of the timber must be clean, smooth and dry, and the manufacturer's instructions should be carefully followed.

4 DESIGN

The design process

The basic process of design is simple, but each step needs careful consideration.

The first step is to identify the problem, and for most construction projects, this is straightforward. *Mother needs a shelf to keep her cookery books on! We need an additional toilet to avoid the problems in the mornings now that the kids are growing up!* But this is an important step, for, by setting out the problem in detail, alternative solutions may come to mind.

These alternatives form the next step in the process, that is, preliminary ideas, and it is here that discussion with other members of the family can help. The best idea is then refined, and a detailed look at the construction methods, materials and time, together with a cost estimate, can then be analysed and a decision made.

The final step is the implementation, but unless the scheme has been thought out beforehand, the actual construction can be difficult. It is far easier to change your mind on a sketch or drawing, than when the wall is half built.

In industry, the construction of several prototypes of a new project is often the norm, but the home carpenter has only one chance; therefore, time spent in detailing out the work beforehand is essential. Having sketched out the ideas, it is sensible to ask a few questions:

Does it solve the problem?
Does it look all right?
Can I actually construct it?
Can I afford it?

In order to answer the first question, it may help to analyse the problem in detail (fig. 34). For example:

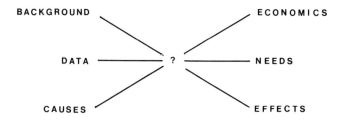

34 Analysing the problem

Background – Having completely redecorated the house an outside activity is needed to balance leisure time.

Data – The area at the rear of the garage is very poor soil, not really suitable for growing anything.

Causes – The garage is getting very cluttered with tools; a new lawn mower is needed, but there is nowhere to store it.

Economics – Having been to the local garden centre an idea of the cost of a small garden shed has been obtained, but none were really suited to the particular location.

Need – Clearly storage for garden tools and lawn mover, but why not an area for potting, or even space for a proper workbench?

Effect – Tidy up the garage to find out exactly what will need to be stored, and then decide on the size and shape of the new shed.

The above example indicates the various factors that can be involved, and, in a project where a financial outlay is envisaged, it is essential that a certain amount of pre-planning takes place.

Freehand sketches are particularly useful in the initial stages as this is an effective way of exploring ideas quickly.

Structural considerations

Once the basic outline has been agreed, then a detailed look at the structure is required.

In order to simplify matters, various conventions have been adopted to show the supports and loads to which each member may be subjected.

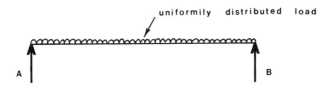

35 Standard beam diagram

Thus, for a simple beam (fig. 35), supported at each end, and with a constant load throughout, a single line is used for the beam. The supports are shown as arrows, and the load is drawn over the entire length. The normal criteria for structural members in the home is not that of failure, but of deflection. Timber, in particular, is a bouncy material and will bend quite dramatically before actually breaking. On a long thin piece of timber, supported at each end, even a small load placed in the middle will produce a noticeable deflection (fig. 36). That same load could be placed on a short chunky piece of timber with no obvious deflection at all (fig. 37).

unloaded

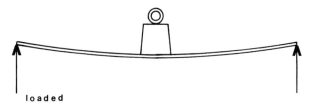

36 *Deflection of a thin section*

loaded

The skill is therefore to find the minimum size that will suit a particular set of loading conditions, yet show little or no deflection. When a beam bends, it is subjected to compressive, tensile and shearing stresses. The top surface is shortened, compressing the fibres together, and is said to be in compression. The bottom surface is stretched – the effect being to pull the fibres apart – and this is said to be in tension. These forces are at a maximum at the top and bottom, gradually decreasing towards the centre, at which point they are balanced out (fig. 38).

Bending strength varies inversely as the length for a similar section, that is if a span of 2m can support a load of 1 tonne, then a span of 4m will only be able to carry a load of a ½ tonne. The span may be reduced to 1m, in which case the load allowed will go to 2 tonnes.

The stiffness, or deflection, varies as the cube of the length, if the section and load are the same. That is; if the span is doubled, then the deflection will increase eightfold; but if the span is halved, then the deflection will only be one eighth.

By varying the width of a beam, the strength will be increased in a direct ratio, the deflection will vary inversely as the width. However, if the depth is increased, then the bending strength will increase directly in proportion to the square, and the deflection will vary inversely as the cube of the depth. In other words, by doubling the depth, the strength is increased four times, and the deflection reduced eight times.

Textbooks are available that cover this area in more detail, but as an

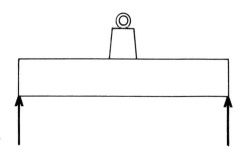

37 *Lack of visible deflection in thick member*

30

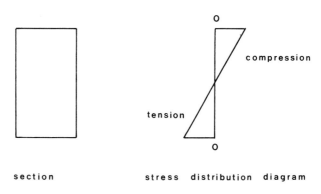

section   stress   distribution   diagram

38 Tension/compression diagram

example of a deflection calculation, the following formula applies for the deflection of a simply supported beam with uniform load:

$$\frac{5}{384} \times \frac{WL^3}{EI}$$

Where:
W = total load on beam
L = distance between supports
E = Young's modulus of elasticity (that is a stress value related to the material)
I = second moment of area, for a rectangle $\frac{bd^3}{12}$

In practice, a deflection of $\frac{1}{360}$ of the span is the maximum that can be tolerated if the beams are to support a plastered ceiling.

In structural design, three factors will need to be considered; the moisture content during use, the size, and the allowable stresses. If timber is to be used in structural alterations to a property, then detailed calculations will need to be submitted for checking and approval before any work is carried out.

However, in order that the home carpenter has some idea of the likely range of timber sizes to be used, some guidelines are given for some typical jobs.

Pitched roof

The rafters give direct support to the roof covering, and may be considered as inclined beams. The size will vary according to the weight of the roof covering; for a light roof they will be either 50 × 75mm (2 × 3in) or 40 × 100mm (1½ × 4in), and for a heavier roof covering from 50 × 100mm (2 × 4in) to 63 × 125mm (2½ × 5in).

Although the spacing will depend on the weight of covering, the centres of rafters are normally between 375 and 450mm (15 to 18in) apart. The

31

wallplate, bedded on to the wall and strapped down, is normally 75 × 115mm (3 × 4½in) to match the course depth and half width of the brick. The ridge which supports the top end of the rafters is generally dependent on the degree of pitch; the steeper the pitch, then the deeper the ridge. Typical sizes would be 25 × 175mm (1 × 7in), 32 × 200mm (1¼ × 8in) or 40 × 225mm (1½ × 9in).

The purlins provide intermediate support for the rafters, and are themselves supported at each end by a wall or truss. They are usually 3 to 4 metres (11 to 15ft) apart, the normal size for a light roof being either 63 × 75mm (2½ × 3in), 75 × 150mm (3 × 6in) or 75 × 175mm (3 × 7in).

Flat roof

For sizing the joists on a flat roof up to 5 metres (16ft) there are rules of thumb which will give an approximate guide. In metric terms: divide the span in decimetres by 2, then add 2, and this will give the depth of joist in centimetres.

For example, a span of 4.2m (that is 42 decimetres) divide by 2 gives 21, add 2 equals 23cm deep. Breadth is normally ⅓ of the depth, so joist size of 74 × 225mm will be the nearest.

In imperial terms: divide the span measured in feet by 2 and call that inches, add a ½ inch for up to 10 foot span, and add 1 inch for spans from 10 to 16ft.

For example, a span of 12ft, divided by 2 gives 6. Call it inches, add 1 as between 10 and 16ft span, to get 7. As before, the width should be about ⅓ of the depth, hence joist size of 7 × 2½in would be appropriate. However, the type of covering, the location and the area should all be taken into account when calculating timber sizes, and it must be emphasized that the methods given are only intended as a guide and are no substitute for proper structural calculations.

Floor joists

The same rule of thumb method may be used to size floor joists when there is timber boarding to be laid. For 18mm (¾in) boards, support should be provided between 300 and 375mm (12 to 15in) and for floor boards of 25 to 30mm (1 to 1¼in) the support should be between 375 to 450mm (15 to 18in).

American practice is generally related to a modular grid to allow the use of 8ft × 4ft plywood sub-flooring, and substantial economies can be achieved with this method by using centre bearing plates to reduce the spans.

5 HOUSE-STRUCTURE

Ground floor

In the nineteenth century and early twentieth century it was common practice in England to construct the ground floor of a house with timber, which would be raised above the level of the ground inside the building. It was essential that the space between the ground and the floor was well ventilated (fig. 39), so that any dampness did not come in direct contact with any of the timber construction.

After the first Model Health Bye-laws were published in 1936, it was compulsory for a continuous layer of concrete 150mm (6in) thick, to be spread between the external walls of the building, thus preventing dampness rising from the ground into the building. Most of the raised timber floors were constructed with imported softwood timber as, prior to the 1940s, this was relatively economical.

The raised timber floor is formed inside the external walls, and supported upon a honeycombed brick sleeper wall, generally from three to six courses high, that is laid directly on to the concrete slab.

A damp-proof course is laid on top of the sleeper wall, with a wallplate bedded on top, usually 100×75mm (4in $\times$ 3in) on to which the floor joists are laid and fixed. Timber boarding is then laid across the joists and skew-nailed to provide a firm level floor.

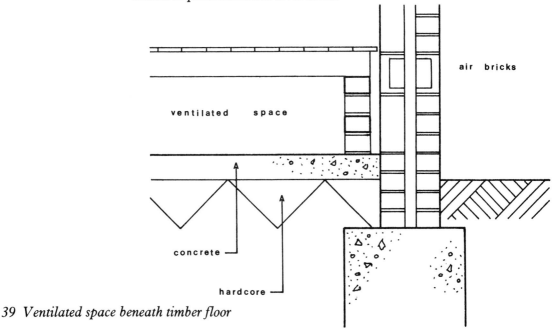

39 Ventilated space beneath timber floor

Floor joists vary in size, but are usually from 38 to 50mm (1½ to 2in) thick, and from 75 to 125mm (3 to 5in) deep, and spaced from 400 to 600mm (16 to 24in) apart. The thickness of the floorboards will be related to the spacing of the joists. For example, a 16mm (⅝in) board will require joists at a maximum of 500mm (20in) between centres, whereas a 28mm (1⅛in) board can have joists spaced as far apart as 790mm (31in) centre to centre. Any length of timber 100mm (4in) or more in width and under 50mm (2in) thickness may be classified as a board, although it is not recommended that floorboards should exceed 125mm (5in) in width.

The edges of floorboards are generally planed to give a projecting tongue on one side, and a groove on the other edge, known as tongued and grooved, or T & G for short (fig. 40). Cramping the boards together as they are laid ensures that the tongues fit firmly into the grooves, and that there are no open cracks between the boards. It is general practice to clamp four boards at a time, and then to nail the boards to the joists using two nails for each fixing.

If replacing floorboards, it is preferable to stagger the end (or heading) joint in as regular a manner as possible, ensuring that the ends of each board have adequate bearing on to the joist (fig. 41).

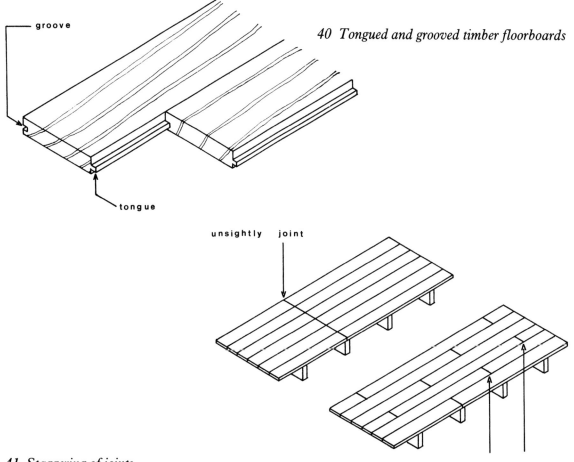

groove

40 Tongued and grooved timber floorboards

tongue

unsightly joint

41 Staggering of joints

staggered joints

34

Upper floor

In order to achieve maximum economy, it has been normal practice for the timber floor joists to span across the least width of any room, with bearing on the external wall and an internal load-bearing partition. The joists will usually be 38 to 50mm (1½ to 2in) thick, from 75 to 225mm (3 to 9in) deep and spaced from 400 to 600mm (16 to 24in) centre to centre.

As timber tends to shrink and warp as it dries out, some form of strutting between the joists is usual to avoid any deformation. The most common type is called herringbone strutting (fig. 42), and consists of short lengths, either of timber or galvanised steel, which are nailed diagonally across the space between the joists. An alternative system is to fix short lengths of timber of the same section in place of the struts, either set in line or staggered (figs 43, 44). A single set of struts would normally be used for spans up to 3.6m (11ft) and two sets for joist exceeding this span.

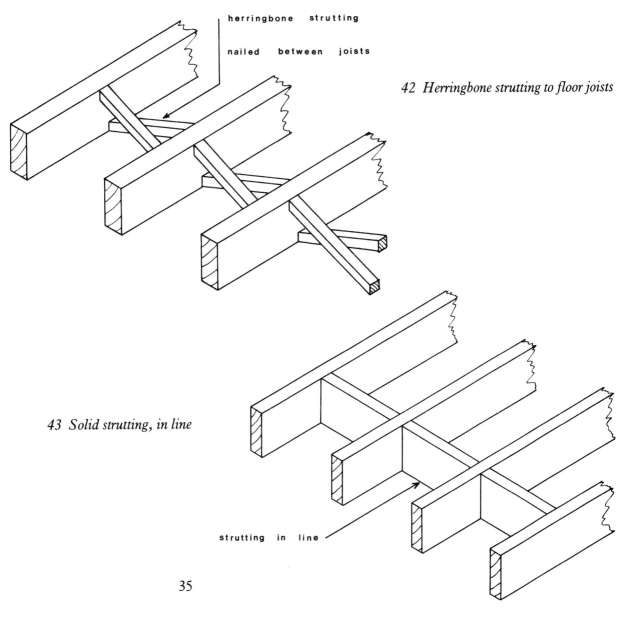

herringbone   strutting

nailed   between   joists

42 Herringbone strutting to floor joists

43 Solid strutting, in line

strutting   in   line

44 Solid strutting, staggered

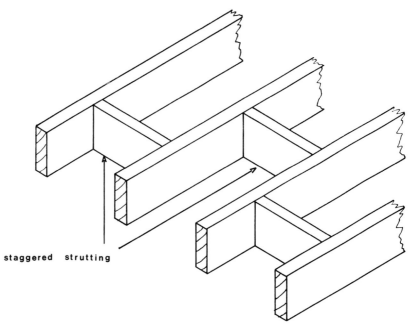

staggered strutting

Timber joists should not be built into external walls, where they could be subjected to damp and gradual deterioration; neither should they be continuous through dividing walls as this could help encourage the spread of fire. The joists should preferably rest on a timber wallplate, which will help to spread the load onto the support wall. Wallplates are normally of 100×65mm ($4 \times 2\frac{1}{2}$in) and should be bedded to level in mortar (fig. 45).

The ends of these joists, as well as the wallplate, should have been treated with some form of timber preservative, particularly if there is any possibility of moisture penetration in the area.

45 Joists resting on wallplate

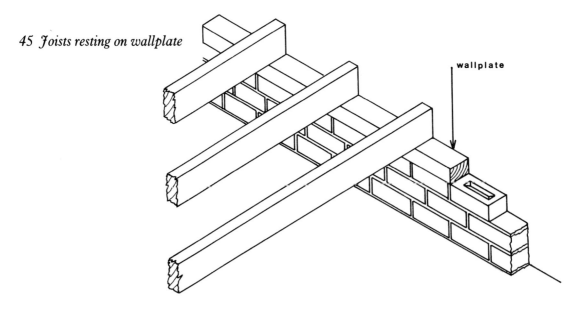

wallplate

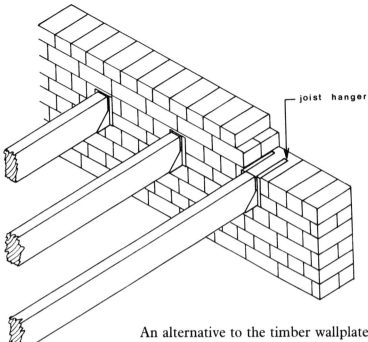

joist hanger

An alternative to the timber wallplate is to use a mild steel bar, 75 × 6mm (3 × ¼in), which should be tarred and sanded before being bedded in mortar. The use of individual tile or slate slips under each joist is not recommended, as it is somewhat laborious and the slips can easily become displaced. The joists can also move during subsequent building operations. Where the timber joists are to be built in a solid external wall, joist hangers may be used to minimize the risk of moisture penetration into the ends of the joists (fig. 46).

As a normal timber-boarded floor with plasterboard ceiling below is a poor sound insulator, it is advisable to install some form of insulation – either fibreglass or polystyrene – between the floor joists (fig. 47).

If fibreglass quilting is laid over the floor joists prior to the nailing of the floorboards, then sound penetration is reduced to a minimum as there is no direct path through the timber.

It has become common practice in both England and America for either high-density chipboard or plywood to be used for flooring. Both materials are available in various standard sizes and grades, but careful planning is

47 *Insulation between floors*

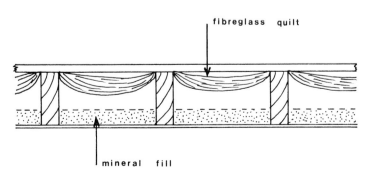

fibreglass quilt

mineral fill

needed if excessive wastage of material is to be avoided. In the case of planks or boards, they should be fixed so they span across the floor joists, with additional noggins used where the edge of the planks join. One major advantage of using chipboard or plywood is that both may be rapidly installed, and usually will ensure a squeak-free floor.

Both materials may also be glued using a 6mm (¼in) bead of glue placed along the top of the joists, with a double bead at the ends. It is claimed that, on a floor with 16mm (⅝in) plywood on 50 × 200mm (2 × 8in) joists, an increase of 25% in stiffness is achieved.

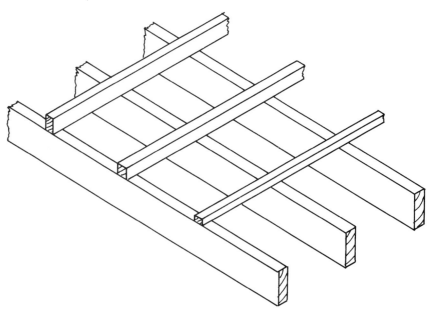

48 Roof slope using differing thickness battens

Flat roofs

Flat roofs are usually constructed so that there is a slight fall towards the rainwater outlets. This can be achieved by laying the joists at a slope (fig. 48), but it is normally done by means of firring-pieces (fig. 49), which consist of tapered lengths of timber, nailed on top of each joist.

The construction of a flat timber roof is very similar to that of the timber floor construction, and, as with the floor, some form of intermediate strutting should be provided. However, a flat timber roof provides poor insulation, and an insulating material – either boards, slabs, quilts or loose fill – should always be included in the design. The weather-proof covering is either a built-up bitumen felt, mastic asphalt, or a non-ferrous sheet metal, such as lead, copper, zinc, or aluminium. All of these materials need careful detailing at joints and edges to be effective and provide reasonable protection.

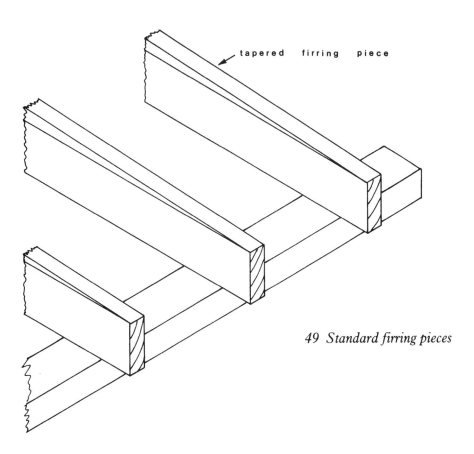

tapered firring piece

49 *Standard firring pieces*

Pitched roof

The most common type of roof is that of the symmetrical pitch to a central ridge, with equal slopes either side. The simplest form, a couple roof (fig. 50), tends to apply unacceptable horizontal forces to the support walls, and it is therefore only suitable for spans up to 3m (10ft) (fig. 51).

It is therefore normal to find ceiling joists nailed across, generally at wall-plate level, to both the tie the roof together and to provide support to the ceiling (fig. 52). A collar roof is a modified close-couple roof, with the ceiling ties fixed at a maximum of one third of the height of the roof.

Today, the majority of timber-framed pitched roofs are constructed as trussed rafters, often prefabricated to a standard design (figs 53, 54). A trussed rafter can use up to 60% less timber than a comparable traditional pitched roof, and will require less labour in erection.

Rafters, ceiling joists and trussed rafters will all be fixed by skew-nailing to the wallplate, but where a lightweight roof covering is used, straps, framing anchors or truss clips are required, as good anchorage is needed to combat the risk of displacement by wind suction.

Various details are used in the construction of eaves, flush or open, closed or sprocketed, but each must allow for some form of ventilation into the roof space (figs 55-58).

39

50 A couple roof

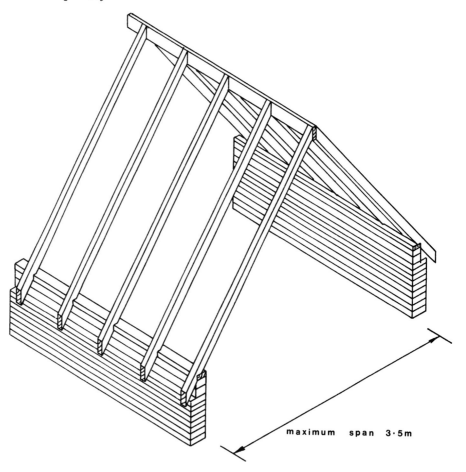

maximum   span   3·5m

51 Forces on a couple roof

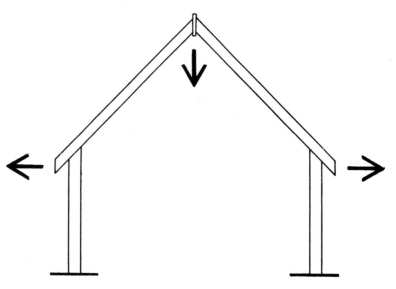

52 A close couple roof

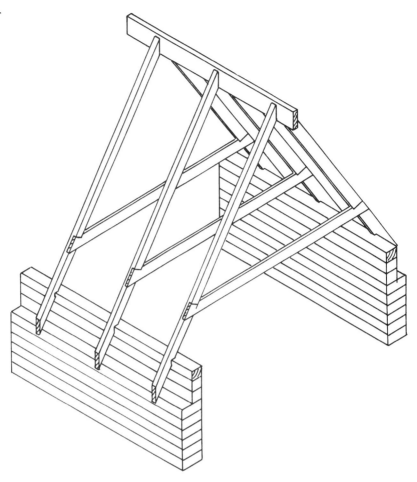

53 A basic roof truss

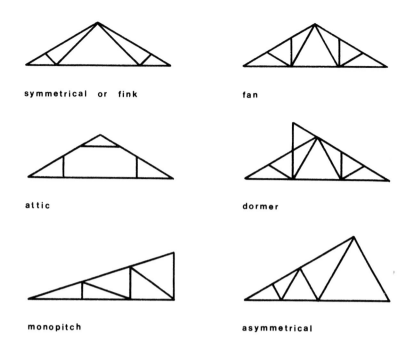

symmetrical or fink

fan

attic

dormer

monopitch

asymmetrical

54 *Standard types of trussed rafters*

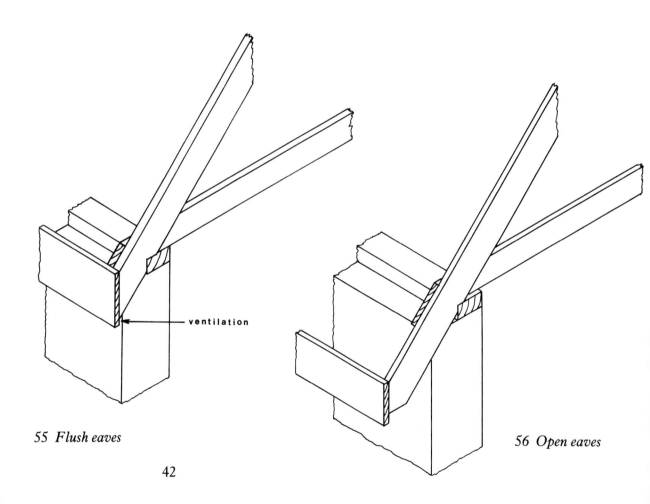

ventilation

55 *Flush eaves*

56 *Open eaves*

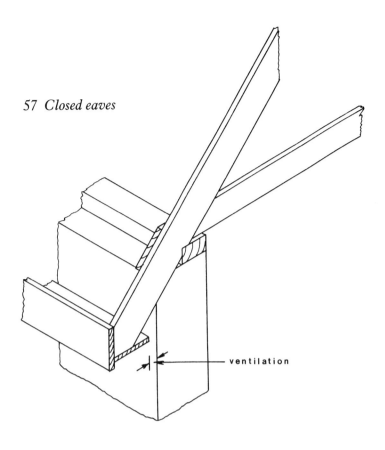

57 *Closed eaves*

ventilation

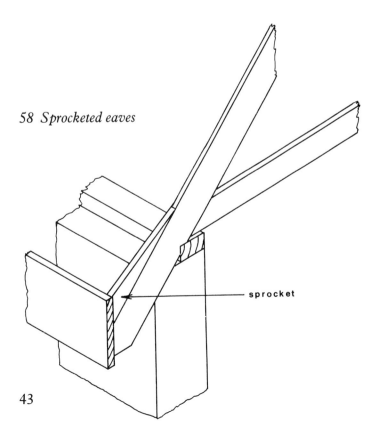

58 *Sprocketed eaves*

sprocket

43

Openings

Wherever an opening is formed in either a timber floor or roof, additional timbers will need to be incorporated to help distribute the variation in loading. This is referred to as trimming (fig. 59), and there are generally two basic requirements for these trimmers; firstly, to trim around an opening, in say the floor (A and B), and secondly, to reduce the span of floor joists over long spans (example C).

For spans up to 2m (6ft 6in) it would normally be possible to use a trimmer consisting of either double or treble floor joists nailed together, but for larger spans, different sections would be required. The most common places to find trimmers in floor construction are either at the stair

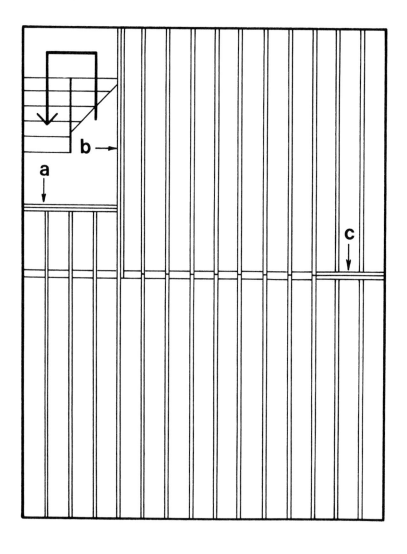

59 Basic types of trimmers

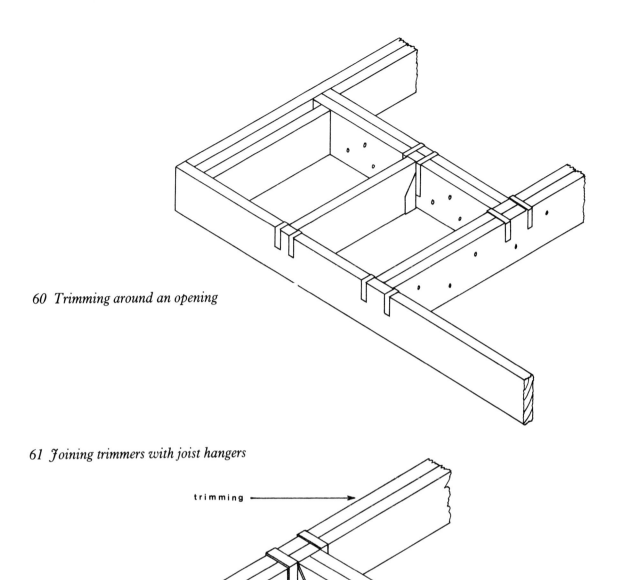

60 Trimming around an opening

61 Joining trimmers with joist hangers

trimming ─────→

trimmer ─────→

well or around a fireplace where the timber joists have to be kept at a minimum distance away from the hearth and chimney to avoid any risk of fire.

Various methods are used to make the different connections involved, some using purpose-made joist hangers, but the more traditional type relies upon carefully cut tenons and housings (figs 60-63). A knowledge of how to construct these connections is essential if a new stair well is being

45

62 Metal connections on roof members

considered. Although it may appear complicated, the work can be completed relatively easily if an investigation of the existing structure is carried out. Load-bearing members must be adequately supported, and the trimmed opening constructed before any attempt is made to install the new stairs (fig. 64).

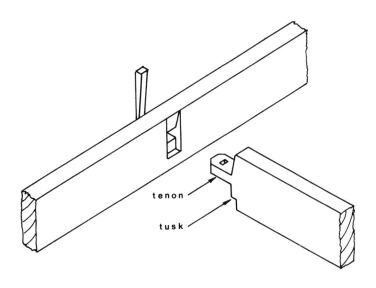

tenon

tusk

63 Traditional joint for trimmer

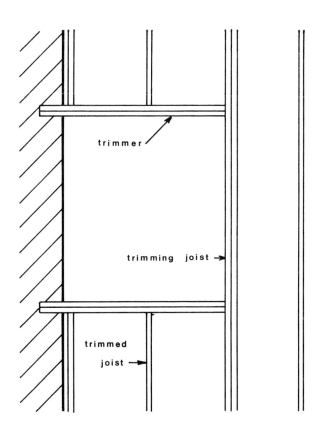

trimmer

trimming joist →

trimmed
joist →

64 Details around an opening

HOUSE - FINISHES

Skirting boards or baseboards

The main function of a skirting board is to provide the transition from wall to floor, and to cover the joint between them. The boards can be machined to a wide variety of profiles, each profile related to the quality and character of the room (fig. 65).

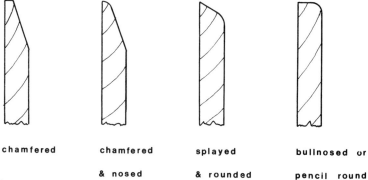

chamfered   chamfered   splayed   bullnosed or

                     & nosed   & rounded   pencil round

65 Standard machined profiles

Skirtings can be made from either hard or soft wood, the latter normally being used if the finished work is to be painted. Fixing of the skirting board to the wall is usually done by one of three methods:

    (a) plugging
    (b) timber grounds
    (c) direct nailing

Plugging (fig. 66), either with wood or plastic plugs, is not always the most satisfactory, as it is often hard to locate the plugs accurately, and no allowance is made for variations in the surface of the plaster.

Timber grounds (fig. 67), although involving more time and effort, do provide a line for the plaster to be laid to, as well as making it easier to fit the skirting board. They also provide a small space behind, that can be used for wiring and so forth.

Direct nailing speaks for itself.

The layout of the room will determine the type of joints required (fig. 68), either

    (a) butt joint – against the architrave
    (b) mitre joint – on an external corner
    (c) scribed joint – at an internal corner

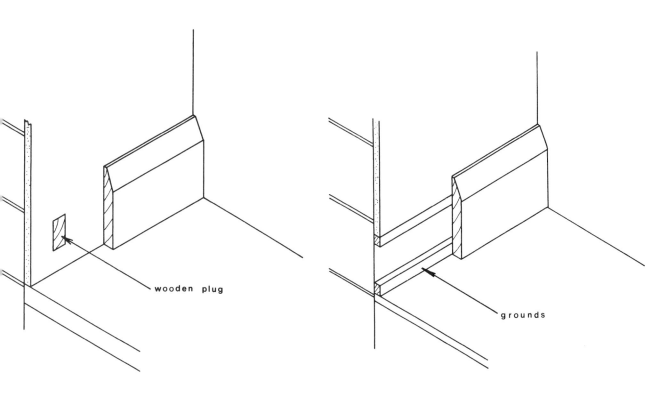

66 *Fixing to wall using plug*

67 *Fixing to wall using grounds*

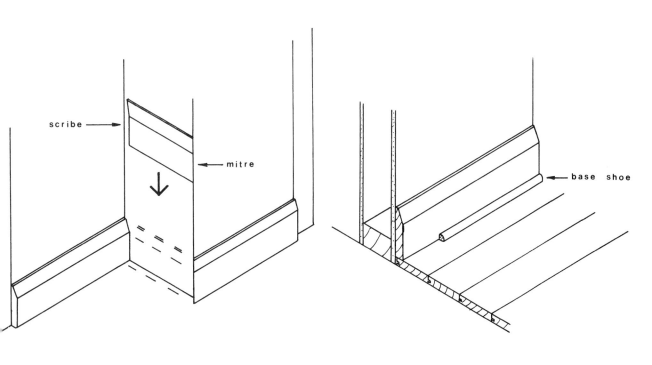

68 *Scribed and mitred joints*

69 *Base shoe, or quadrant*

49

It is not advisable to join skirting boards lengthways, as the normal shrinkage of the board will expose the joint. Base shoe (or quadrant) beading (fig. 69) is not particularly popular in England, but may be seen as a seal between the baseboard and the floor finish.

Architraves, trim and casing

Architraves are the decorative surround to any opening. They often cover the junction between the wall surface and the frame within the opening. In older houses, architraves are often elaborate mouldings, but in modern construction they are generally of a much simpler section.

When the older mouldings are damaged, it is often difficult to find an exact replacement, and therefore alternatives have to be considered. These are:

(a) replace the entire length of old moulding with a modern moulding
(b) find a similar moulding elsewhere that could be replaced without being noticed
(c) make a new section to exactly match the exisiting one.

The last alternative may require the use of an 'instant shape tracer' made out of small sliding metal rods set in a clamp. Then plane up the required shape using the various combinations of moulding blades available.

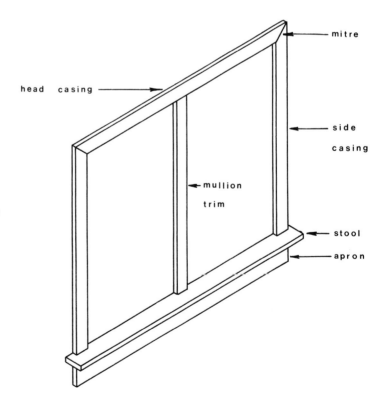

70 *Trim around a window*

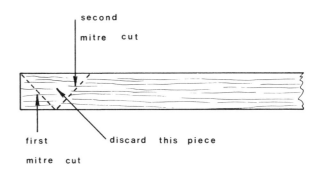

second

mitre   cut

first

mitre   cut

discard   this   piece

71 An end mitre

It may be possible to combine several modern sections and actually build up a similar section to that needed. In timber-frame houses, it is normal for trim to be provided around a window opening, using a casing, stool, apron and stops (fig. 70). These are American terms, although, in modern practice, the stool and apron may be replaced with a bevelled sill liner and casing. This style is referred to as 'picture frame' trimming. In order to avoid end grain showing, the appearance of the apron may be improved by cutting two mitres, discarding the triangular piece and glueing the end piece back at 90° to the main section (fig. 71).

Panelling

In many good-quality period houses, timber panelling is a major feature, often with inlays or carving on the panels. Modern practice is to provide a timber framework of battening attached to the wall, and then to fix the boards or panels to this.

The use of a room, its size, shape and lighting will all influence the design of any panelling, although personal taste in deciding whether to line one wall or the whole room will be the deciding factor.

If the wall to be panelled has any tendency to dampness, then the timber should be treated with a damp-proofing compound, or some form of vertical damp-proof membrane, such as polythene sheeting, aluminium foil or heavy duty bitumen-impregnated paper should be considered.

Timber boarding should be kept in the same room for at least a week, in order to allow the excess moisture to evaporate, for it to take on the same humidity as the room.

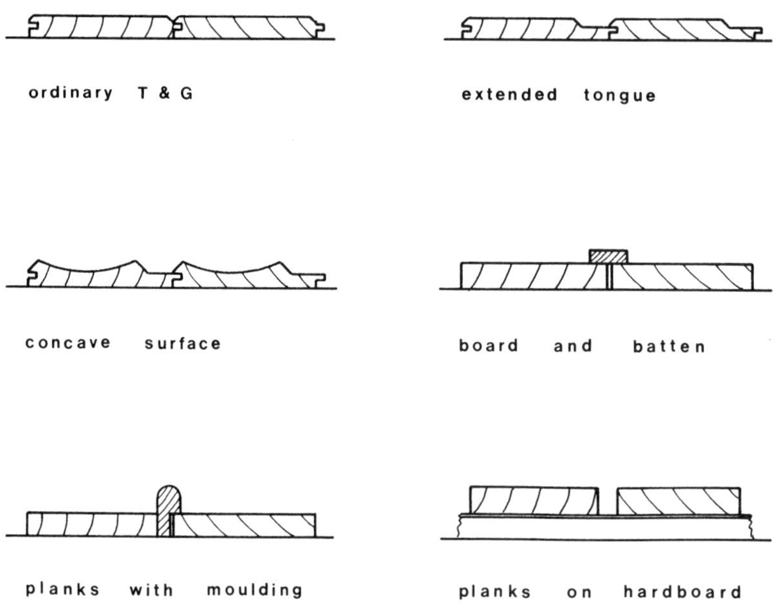

ordinary   T & G

extended   tongue

concave   surface

board   and   batten

planks   with   moulding

planks   on   hardboard

72 *Types of panelling*

Various sections of timber can be used to line a wall (fig. 72). An unusual and attractive variation for a plain wall is to place the boards diagonally instead of the more usual horizontal or vertical.

Careful planning must always be the rule when working around doors and windows, as diagonal boards are not always appropriate in this situation. Large sheets of veneered plywood are often used, again being fitted to timber battens nailed to the wall. The positioning of the battens will depend on the type of boarding being used, and will run in the opposite direction, either vertically or horizontally, to the finished boarding.

Battens can be fixed in a variety of ways. For masonry walls some form of plugging and screwing is probably the most satisfactory. For timber walls there is less problem as they can be nailed direct to the main studwork.

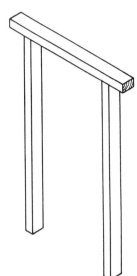

7 DOORS

Frames

Door frames are needed to provide a fixing for doors, and therefore the sections of timber used must be large enough to enable the frame to remain rigid, especially the jambs on to which the doors will be hung.

Door frames can be broadly classified as internal or external (fig. 73), the four main types being:

    (a) door frame with jambs and head
    (b) door frame with jambs, head and threshold
    (c) door frame with fixed fanlight, consisting of jambs, head and transom.
    (d) door frame with hung fanlight, consisting of jambs, head and transom, threshold and opening fanlight.

Combined window and door frames (figs 74, 75) are often used for residential houses, both door and casement usually being designed for outward opening, although the door frame may be for inward opening.

ngle   door   frame

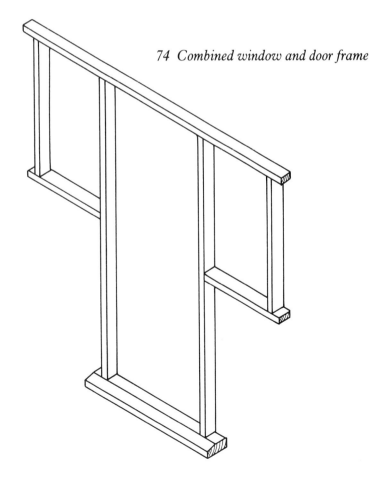

74 Combined window and door frame

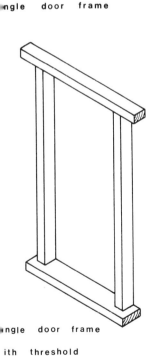

ngle door frame

ith   threshold

73 Single door frames

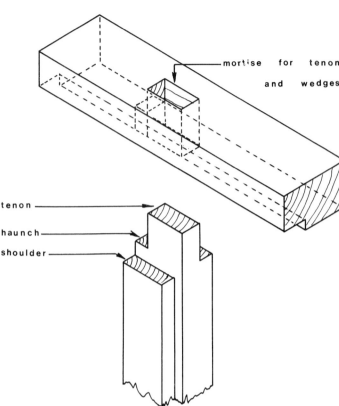

mortise for tenon and wedges

tenon

haunch

shoulder

75 Combined window and door   *76 Detail of top mortise-and-tenon joint*

French windows, used in Europe, open inwards and lead to either a balcony, verandah or terrace. In other countries, the opening may be outward, but in most places they are now being superseded by aluminium-framed sliding patio doors.

In all good-class joinery, the joint most commonly used in the construction of door frames is the mortise and tenon (fig. 76), secured by one of the following methods:

(a) draw pinning
(b) wedging and pinning
(c) nailing

Draw pinning has the advantage that the joint can be pulled up without the use of cramps, but it is not suitable for modern methods of mass-production. With wedging and pinning, the joints may be glued and pinned, or glued and then wedged.

Standard door frames are available from most builder's merchants and timber suppliers, but for non-standard sizes it is better for the home carpenter to make his own.

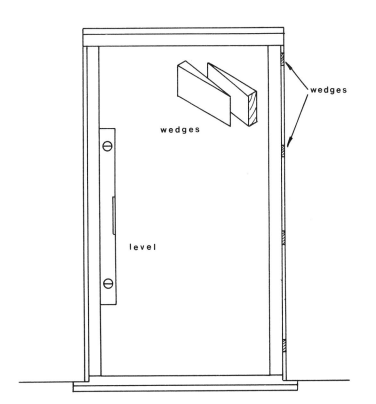

77 *Use of wedges to set frame vertical*

To install a new door frame is a relatively simple task, particularly if the threshold or sill is attached. Place the frame in the centre of the opening, and secure it with a temporary brace. Using blocks and wedges, level the threshold, and fix it at the required height. Then, insert blocks and wedges at the bottom, centre and top between the jamb and the side of the opening, ensuring that the jambs are vertical (fig. 77).

These wedges can be adjusted so that the side jambs are well supported, and then the wedges can be secured by driving a nail through the jamb and wedges into the supporting wall.

Door frames that do not have a threshold or sill can be secured to the floor by a mild steel dowel rod, 12mm (½in) in diameter, and some 50mm (2in) long, that is driven into the base of the jamb and set into the floor.

The traditional method of excluding wind and rain with outward opening doors relies on the rebates in the door frame, with the addition of a weatherboard (fig. 78). However, the addition of compression seals and generous drainage channels in the frame and door can help in exposed conditions (fig. 79).

For internal doors, the jambs or linings are of a thinner section, with the head and side joined together with a simple grooved joint. In America, the normal practice is for the side jamb to be dadoed (or grooved) to receive the head, whereas the English practice is for the top of the side jamb to be tongued to fit into the head lining, which is suitable grooved (fig. 80).

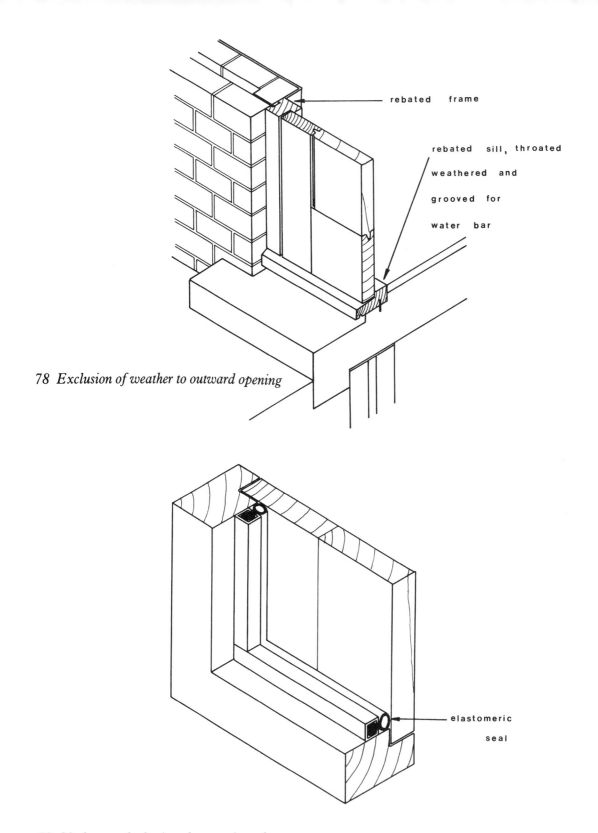

rebated   frame

rebated sill, throated

weathered and

grooved for

water bar

78 *Exclusion of weather to outward opening*

elastomeric

seal

79 *Modern method using elastomeric seals*

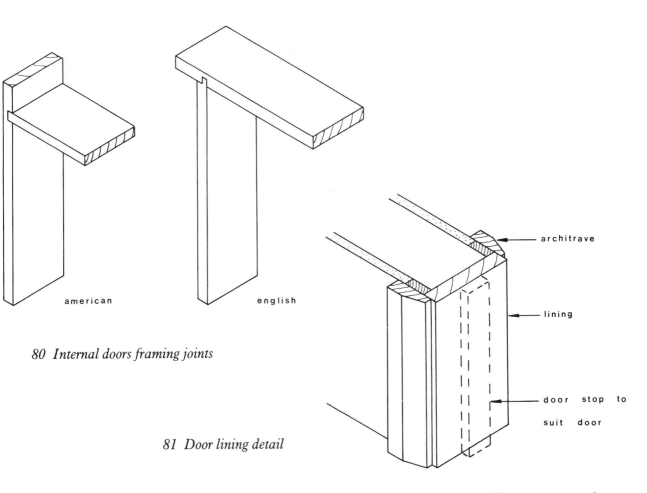

architrave

lining

door stop to
suit door

80 Internal doors framing joints

american

english

81 Door lining detail

Traditional door linings may be single or double rebated linings, or a panel lining fixed to grounds (fig. 81).

Doors

The simplest form of door has a facing of tongued, grooved and V-jointed boards. These may be fixed vertically to either ledges, ledges and braces, or to a frame which is braced, to form a matchboard door. These doors are suitable for cellars, sheds and stores, where the appearance, thermal and sound insulation are of secondary importance.

A ledged matchboard door (fig. 82) has the vertical tongued and grooved boarding nailed to three horizontal battens, or ledges, but this design has the tendency to droop rapidly and fall out of square. It is therefore only suitable for use in narrow openings. A ledged and braced door has the addition of diagonal braces, which help to support and hold the door in shape (fig. 83).

A framed and braced matchboard door (fig. 84) has the boarding fixed to a frame of stiles and rails that are generally joined together with mortise-and-tenon joints, and with the addition of braces to strengthen the door

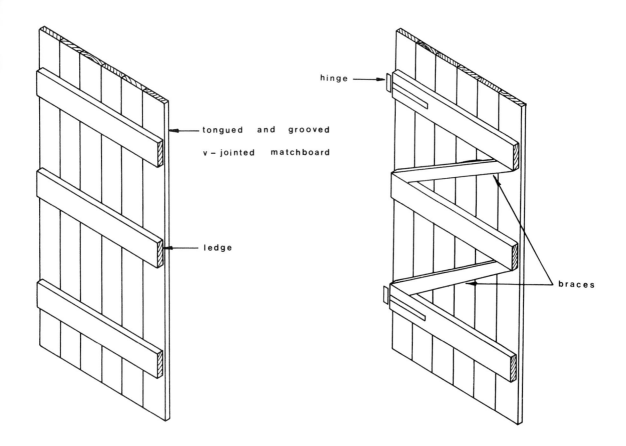

82 Tongued and grooved ledger door　　　*83 Tongued and grooved ledger and braced door*

and prevent it sinking out of shape (fig. 85). This latter type of door can be used for large openings, such as garages or entrance gates. Provided care is taken in the setting out of the various members, the construction of a framed and braced door (fig. 86) can give a great deal of satisfaction. There are many different styles of doors, but broadly they fit into two main categories, panelled or glazed (figs. 87-88). The classic panelled door consists of a frame made up of stiles, rails, and muntins, with the panels grooved into the frame, and with the framing moulded out of the solid. Bead-and-butt panelled doors have the panels tongued into the frame, but with a bead on the vertical edge only (fig. 89).

For the construction of the main frame, two types of jointing are commonly used: the mortise, tenon and wedge, or dowel construction. Internal doors may be flush, panelled or glazed. Modern production methods now allow the doors to have the panels pressed out of hardboard, or composition board, and although much lighter in construction than solid timber doors, when painted, they present a similar appearance (fig. 90).

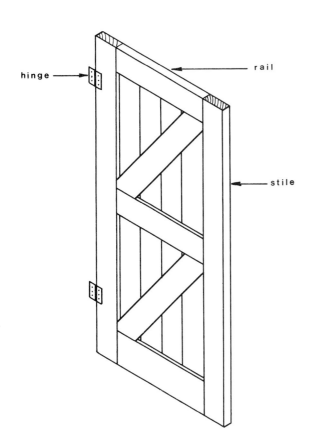

84 *Framed and braced matchboard door*   85 *Framed and braced door incorrectly hung*

Flush doors may have a solid or framed construction, but, unless the opening is of a standard size, the latter may not be suitable, as a reduction in either height or width may considerably reduce the cross-sectional area of the frame, and hence weaken the door.

Fitting doors

Doors will invariably change shape very slightly, depending on the temperature and humidity of their surroundings. Even though this effect will be small if they are correctly painted or varnished, it must be taken into account at the time of fitting. The normal clearance for an internal door is around 3mm (⅛in) on the top and sides, and 6mm (¼in) at the bottom. It is essential that the door to be fitted is left in the room for two or three days so that it can adjust to the prevailing humidity.

The first step is then to offer up the door to the hinge side of the door frame, and plane that edge so that it matches the frame. After this, do the

59

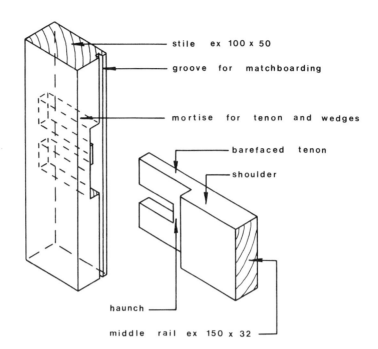

stile   ex 100 x 50

groove   for   matchboarding

mortise   for   tenon   and   wedges

barefaced   tenon

shoulder

haunch

middle   rail   ex   150 x 32

86 Detail of typical joint

87 Pair of glazed doors

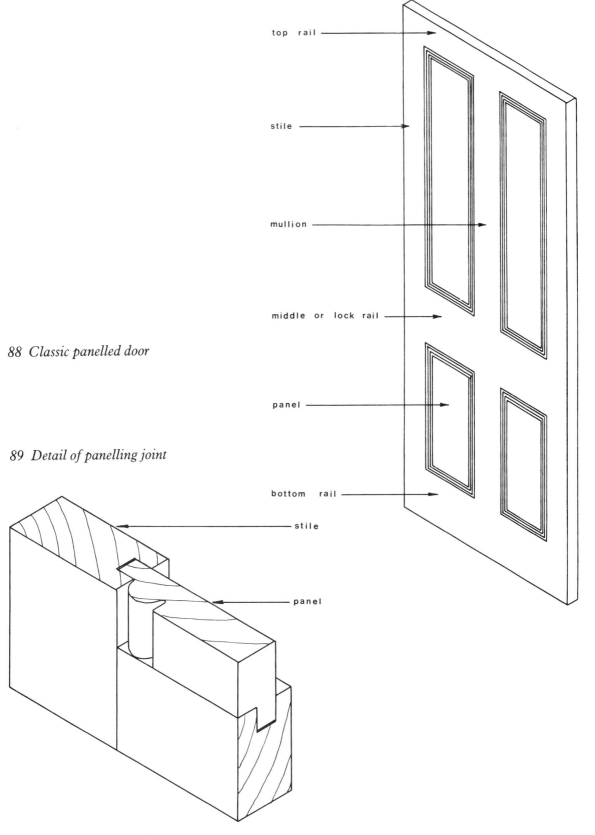

top rail

stile

mullion

middle or lock rail

88 Classic panelled door

panel

89 Detail of panelling joint

bottom rail

stile

panel

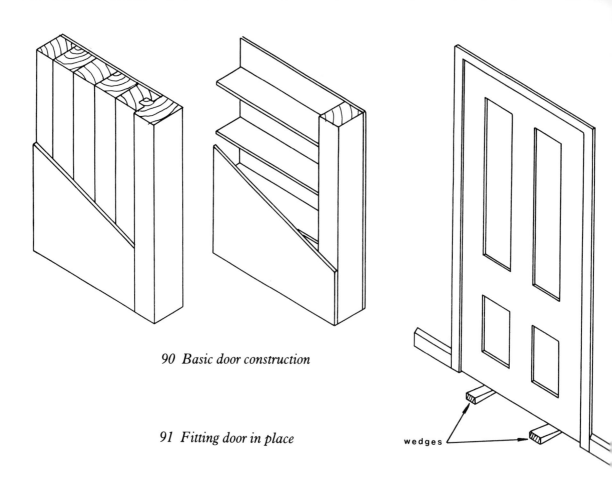

90 *Basic door construction*

91 *Fitting door in place*

wedges

same for the top, and then the bottom, and finally the side which will house the lock and handles.

To fit the hinges, the door must be placed in position so that the door and frame can be marked at the same time. The door should be held firmly in place by the careful use of wedges (fig. 91). Having done this, mark down 150mm (6in) from the top and 225mm (9in) up from the bottom, marks being made at these levels on both door and frame.

The door is then taken down, and the position of the hinges is carefully indicated with a marking knife, the hinges being positioned on the inside of the initial marks. Using a broad chisel, these areas should then be cut out so that the hinges fit neatly into the recesses (fig. 92). With the hinge in position, drill one guide hole for one screw in each flap, and put the one screw in each part.

Gently open and shut the door, making sure that it fits flat into the frame, and that it stays there without having to be held. It may be necessary to put a slight bevel on the lock side of the door in order to avoid sticking. Also, check that the door can open fully, and that it will not catch on the floor or carpet. When satisfied, the remaining guide holes can be drilled and the countersunk screws carefully screwed into place.

The handle, lock and catchplate will generally come as a complete package, with detailed instructions from the manufacturer relating to installation. It is essential to read through these instructions before commencing to fit the ironmongery.

It is possible to buy a pre-hung door unit which consists of a door frame with the door already installed, complete with handle, lock and key. This type of unit is really only suited for installations when the opening is reasonably accurately made, for example within a timber-framed wall.

Sliding doors allow a space-saving feature, as the door doesn't open into the room, but slides, either into the wall, known as a pocket type, or against the wall – the latter being easiest to install in an existing house.

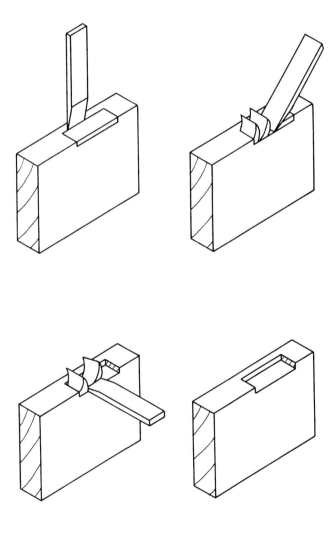

92 *Cutting rebates for hinges*

WINDOWS

Windows may be classified under one, or a combination of three basic types:

- (a) fixed
- (b) swinging
- (c) sliding

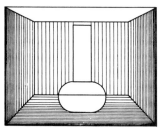

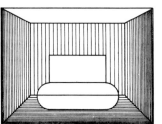

93 Quality of illumination related to shape

The main purpose of any window is to allow light into a building, and to be able to exclude wind, rain and snow. In addition, it can serve as a means of providing ventilation if the glazed area can be opened, and can also allow the occupants to have a view outside. The quality of natural lighting in a room will generally depend upon the size and shape of the window in relation to the area of the room, upon the outside orientation, and any possible obstructions that may keep the window in shadow (fig. 93).

Calculation of required window size can involve the use of a 'daylight factor', which is the ratio of internal illumination to the illumination occurring simultaneously out of doors from an unobstructed sky. This ratio varies from 0.5% to 6%, and is roughly equal to one fifth of the percentage ratio of glass to floor area. For example, in a room 3m × 4m (12ft × 15ft), where a daylight factor of 4% is required, the area of glass will need to be

$$4 \times (3 \times 4) \times \frac{5}{100} = 2.4 m^2$$

That is, about 20% of the floor area (20% of 12ft × 15ft will give around 36ft^2)

For general comfort, a natural change of air to ventilate a room is desired, and a figure of ½0th of the floor area as ventilation openings is acceptable. Ventilation is necessary to minimise condensation, as it allows warm moist air to be exchanged for cooler dryer air.

A window must be strong enough to resist the pressures and suction due to the wind, as well as being securely fixed in the wall opening. It must also be considered in relationship to thermal and sound insulation.

Fixed light

A fixed light or window is where a sheet of glass is fixed, either directly or indirectly, to the wall structure, often into a frame – which, in turn, is fixed to the wall – so that no part of the window will open (fig. 94).

64

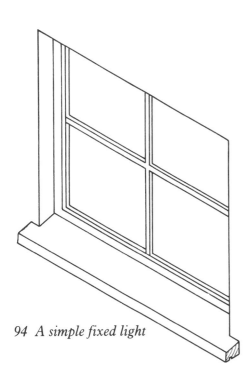

94 A simple fixed light

95 Fixed lights

96 An opening light

Fixed lights are two or more panes fixed to the frame with intermediate supports – such as mullions, transoms or glazing bars – all fixed so that nothing opens (fig. 95).

Opening light

An opening light or window is the part that can be opened by being hinged or pivoted to the frame, or can slide open inside the frame (fig. 96).

They can be further classified as:

    (a) side, top or bottom hung
    (b) horizontally or vertically pivoted
    (c) vertically or horizontally sliding.

Side hung casement windows

For centuries, wooden casement windows have been the traditional form of windows for small buildings and even now, most timber manufacturers include a wide ranges of sizes and styles of this type (fig. 97).

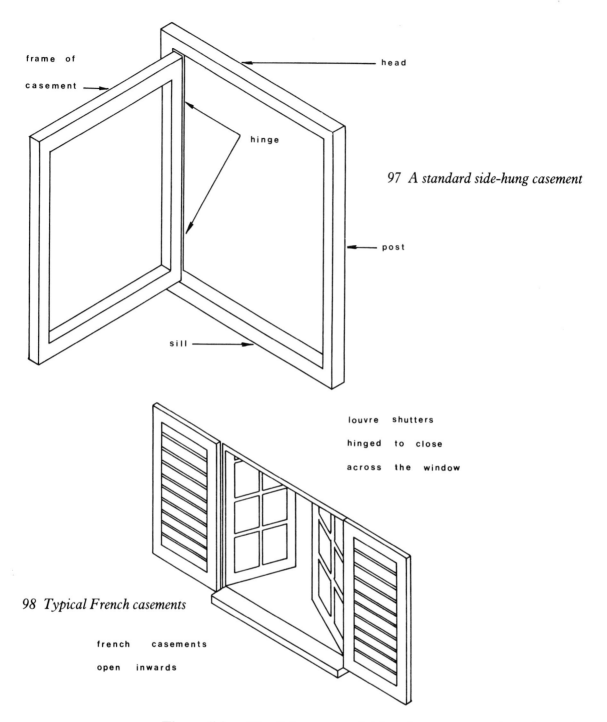

frame of
casement

head

hinge

97 *A standard side-hung casement*

post

sill

louvre   shutters
hinged   to   close
across   the   window

98 *Typical French casements*

french   casements
open   inwards

The traditional English casement is hinged to open out as it can be made to exclude wind and rain more easily than in inward opening window. However, elsewhere in Europe, the traditional method has been for casements, generally made in pairs, to open inwards with a set of outside louvres that can be closed when the casement is open to give ventilation and shade from the sun (fig. 98).

66

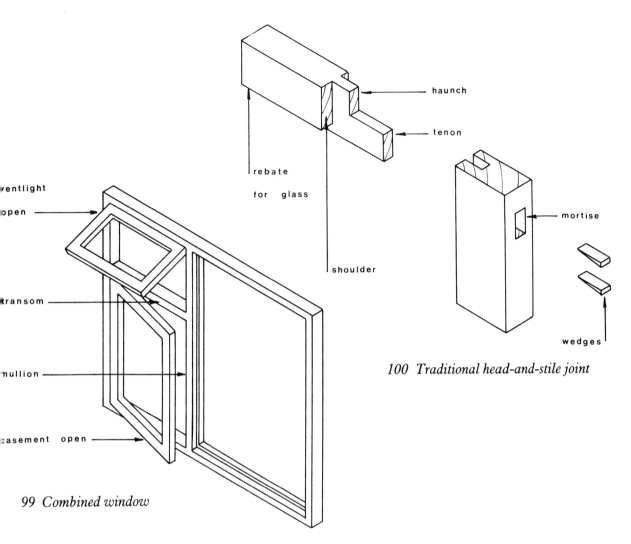

ventlight
open

transom

mullion

casement open

99 *Combined window*

haunch

tenon

rebate
for glass

shoulder

mortise

wedges

100 *Traditional head-and-stile joint*

Because a casement is hinged on one side, there is a tendency for the other side to drop (due to the weight of the glass and timber frame) and this will result in the window binding in the frame. It is therefore unwise to make a casement wider than 600mm (2ft) and, although a central mullion will obstruct the view, it is a much sounder design to include it (fig. 99). One disadvantage of outward opening casements is the difficulty of cleaning from the inside. Replacement of a casement window is often necessary as the original joints may weaken, with the result that the frame will tend to sag.

The craft of accurately cutting and joining the timber members of windows and doors gave rise to the term 'joinery', and although modern practice is to use machinery for this work, the construction of an individual casement can easily be done by hand. A casement frame consists of a head, two stiles and a bottom rail, traditionally joined together with a wedged and haunched mortise and tenon (fig. 100). In modern construction, the casement frame is made using a combed joint. This is much more suited to mass-production methods (fig. 101).

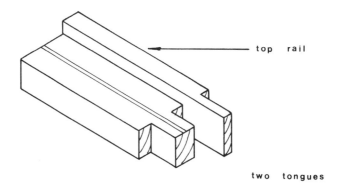

top rail

two tongues

101 Modern combed joint

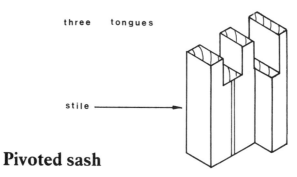

three tongues

stile

Pivoted sash

The sashes may be either horizontally or vertically pivoted to open, the horizontal usually being pivoted at the centre to balance the weight (fig. 102), the vertical being pivoted at the one-third point to facilitate cleaning. However, close control of ventilation with these windows is not possible, as they have to open both top and bottom, or both sides, and can catch and direct gusts of wind into the building.

102 Horizontally pivoted sash

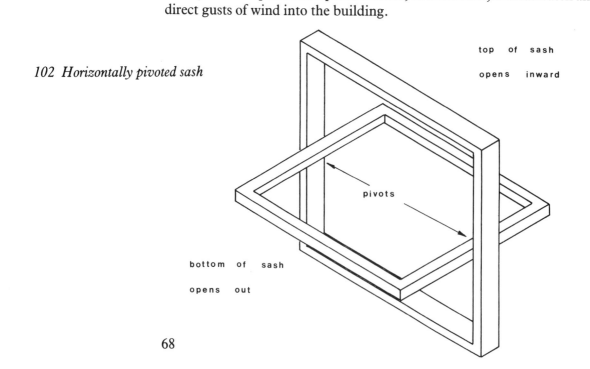

top of sash

opens inward

pivots

bottom of sash

opens out

Sliding windows

Timber-framed vertically-sliding sash windows have the advantage that the sashes are hung, and therefore have less tendency to distort their shape (fig. 103). A major disadvantage is that this type of window is difficult to clean from inside the building.

Renewing the sash-cords on these windows requires a great deal of patience. The beading has to be removed and the sash lifted out in order to relieve the load on the sash-cords.

A small access hatch is provided at the bottom of the pulley stile so that the old cord can be withdrawn and replaced with a new one. It is advisable to use a nylon fibre and flax cord when carrying out this task, as it will have a considerably longer life than ordinary sisal cords.

Another traditional form is that known as the Yorkshire sliding sash (fig. 104), where the two timber framed sashes slide horizontally between wooden beading set in a timber frame.

Although attractive and simple in design, these windows are liable to jam, and can be difficult to open and close.

103 Vertical sliding sash

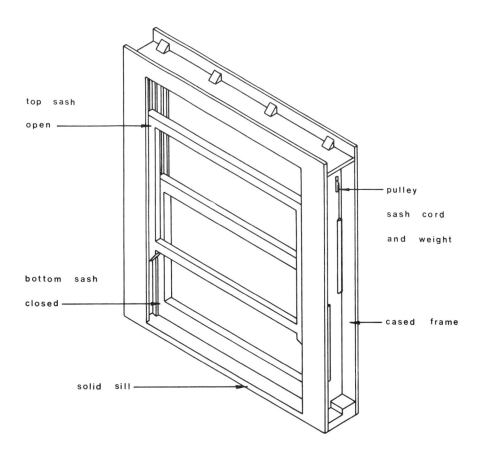

top sash

open

pulley

sash cord

and weight

bottom sash

closed

cased frame

solid sill

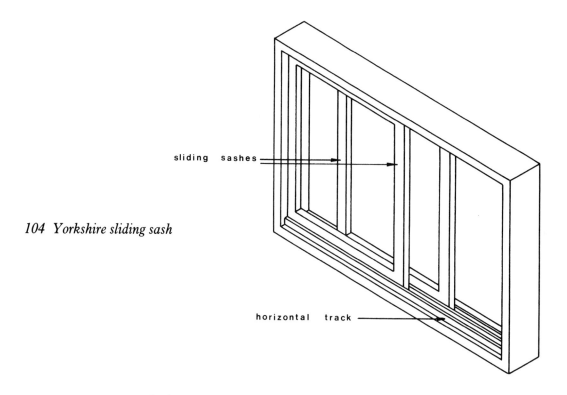

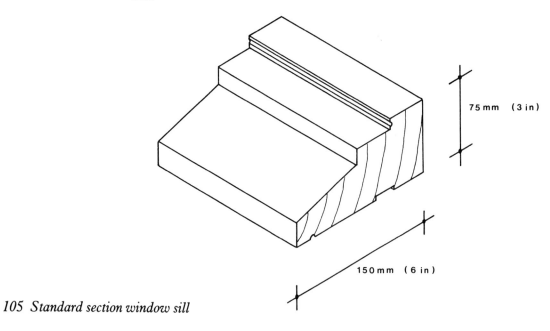

104 Yorkshire sliding sash

Window sills

The basic function of an external window sill is to protect the wall beneath the window from rain (fig. 105). It is normal for windows to be set back from the outside wall, so that they have some protection from the elements, but, as a result, care has to be taken with the design of a suitable sill.

105 Standard section window sill

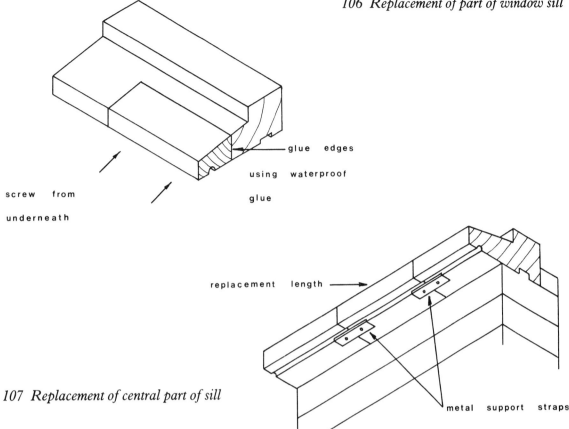

glue   edges

using   waterproof

glue

screw   from

underneath

replacement   length

107 Replacement of central part of sill

metal   support   straps

In the older houses, it often becomes necessary to replace parts of the sill and nearby framework, and this should be done as soon as any deterioration of the timber becomes apparent (fig. 106).

It is essential to cut out and remove any rotten timber so that only sound wood remains. From direct measurements, it is then possible to cut out replacement sections, taking care to provide corresponding drip grooves to match those of the existing sill.

It is generally easier to complete the final shaping and matching once the replacement wood has been fixed, making sure that any countersunk screws used are well below the finished surface (fig. 107).

Before finally fixing the replacement part, a liberal coating of wood preservative is required. The contact areas should be well coated with a water-resistant adhesive, and holes drilled to take the screws. In order to avoid recurring problems, it is important to find out why the damage has happened, and to take any remedial steps required to make sure it will not happen again.

With frame, casement and sill it is essential that there is an efficient drip groove throughout (fig. 108). It may well be that continual painting has reduced its effectiveness, and it may need reforming.

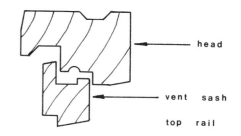

head

vent   sash

top   rail

casement   sash

bottom   rail

sill

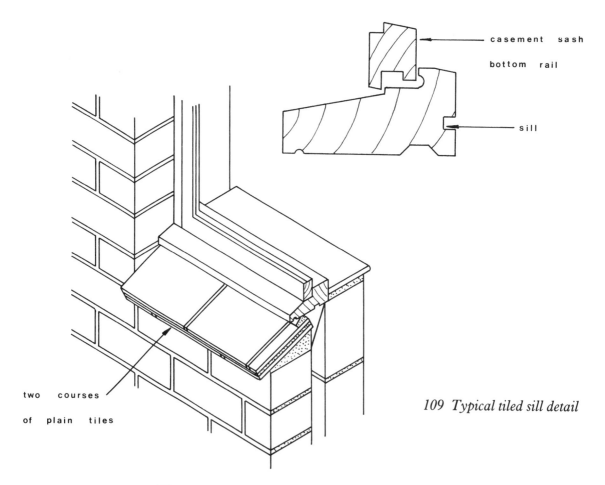

two   courses
of   plain   tiles

109 Typical tiled sill detail

The slope of the sill should be such that the water is quickly allowed to run away (fig. 109). If rot has gone too far, complete replacement may be the easiest solution, and it is then that any major changes may be carried out to avoid future problems.

Prevention, although a lesser challenge than new work, is much better than cure – so any replacement work must be protected and maintained to ensure as little problem for the future as possible.

PARTITIONS

The advantages of constructing a timber-framed partition are that it is quick, clean, and a dry process. The timber may be cut and assembled, either flat, or in position, and, if correctly designed, can give strength and support to the floor and ceiling.

General details

A traditional timber partition has vertical studs joined on to a head at the top, and a cill, or soleplate, at the base, and with noggin pieces added for rigidity (fig. 110). During construction, various features and details can be added: for example, openings for service ducts, or support pieces for plumbing or electrical fittings (figs 111-114).

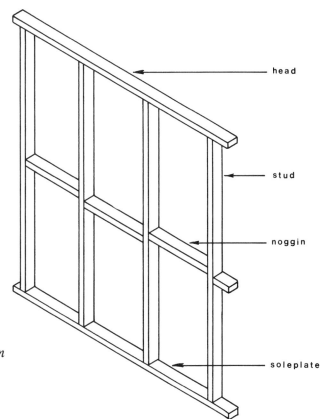

head

stud

noggin

soleplate

110 Basic structure of a partition

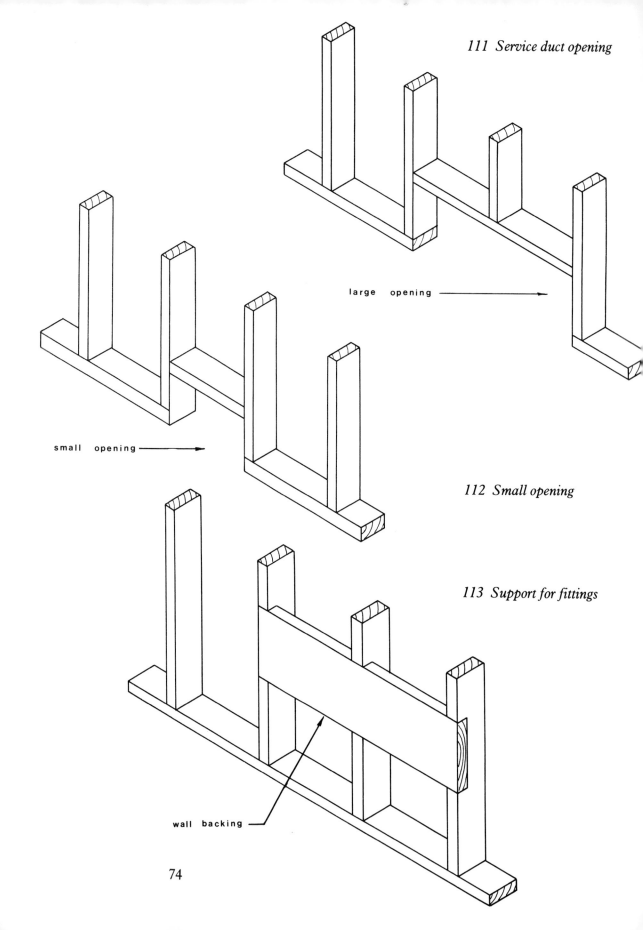

111 Service duct opening

large opening ⟶

small opening ⟶

112 Small opening

113 Support for fittings

wall backing

74

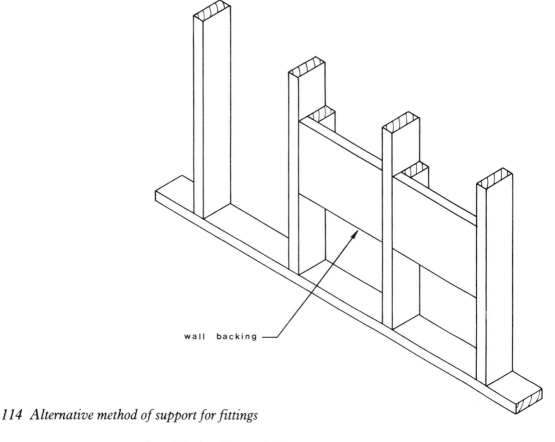

wall backing

114 Alternative method of support for fittings

Careful detailing of the corners is required to ensure that there is sufficient bearing for the wall finishes to be fixed (fig. 115). This is often done by simply providing an additional stud at the corner. One problem with the traditional stud partition is the difficulty in providing adequate sound-proofing. However, if the studs are staggered, then the direct path from one side to the other can be avoided, and an insulation barrier can also be incorporated (figs 116-117). Many different types of material can be used to provide sound insulation, including fibreglass, fibre board, straw or felt.

The vertical studs may be fixed to the head and cill by either a simple butt-joint, trenched or mortised-and-tenoned together (fig. 118).

If the frame can be constructed in a workshop, then it is preferable to use the mortise and tenon joint, but if it is to be constructed *in situ*, then the trenching method, that is, cutting a housing in the head and cill, is the more usual.

The noggin pieces are cut to fit tightly between the studs, and are neither trenched nor tenoned. When considering the construction of a timber stud partition, it is essential that there is adequate fixing available for both head and cill. It is preferable that the cill span across the floor joists, if not, then it must be positioned directly above a floor joist to avoid undue loading on the floorboards (fig. 119).

75

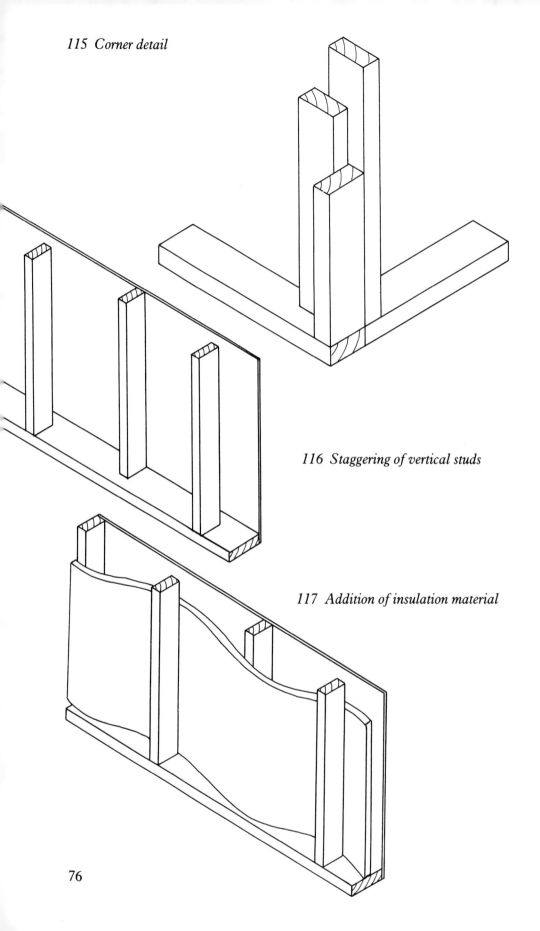

115 *Corner detail*

116 *Staggering of vertical studs*

117 *Addition of insulation material*

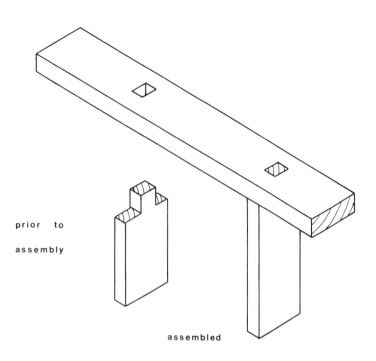

118 Detail of head to stud joint

prior to

assembly

assembled

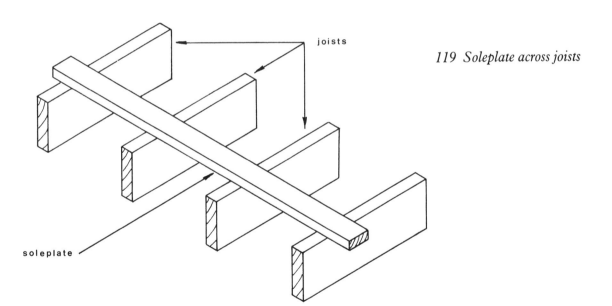

joists

119 Soleplate across joists

soleplate

   In older properties, it is unlikely that the floor will be level, and it may be necessary to pack between the floor and the cill to bring the cill level. If this is not done, then the length of each vertical stud will be slightly different and will have to be seperately measured and cut. Similarly, the head should bear across the ceiling joists, if not, then it may be necessary to cut out the plasterboard, or ceiling board, and install short struts between the joists to give some form of support to the fixing of the head.

Typical installation

If it is proposed to divide a large bedroom into two smaller ones, then the construction of a partition may well have to include a doorway, and perhaps allow for the construction of storage cupboards or even a worktop (fig. 120). These factors must be taken into account in the design stage, and appropriate dimensions used to allow for individual door and frame sizes, as well as the space requirements for the storage unit.

The use of standard-size doors for the fronts of cupboards will greatly speed up the construction time of any project, but it is essential that the sizes are known before the construction and erection of the partition.

In the case of built-in cupboards, the location of shelves and rails should be decided at the design stage, so that any additional bearers can be correctly installed during the basic construction (fig. 121). Generally some form of lining board will be installed. Care must be taken to ensure that the additional structural features are noted so that fixing of the rails and shelves can make use of their support. The timber used for the main elements of a partition must be straight and true, although it is not necessary for it to be planed.

It is often difficult to find sawn timber that conforms to the standard required, and it may be necessary to spend a little more money to get timber that has been planed and is reasonably straight. The difference in cost should not be excessive, and there is always the added benefit that it is more enjoyable to work with a smooth surface than one which is liable to produce splinters.

With the construction of a doorway, it is advisable to strengthen the studs that will form the door jambs, either by using a larger section, or by doubling up the studs. This will then need less strutting to give a vertical edge, as well as providing a more solid background support for hinges and fittings (figs 122-123).

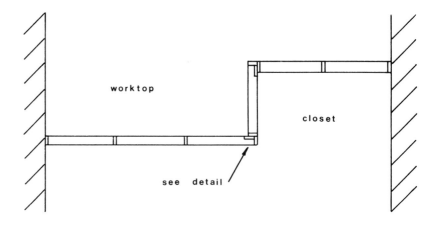

120 Typical plan for partition

78

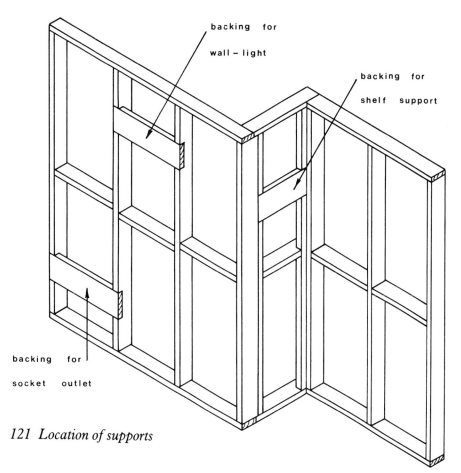

backing for
wall – light

backing for
shelf support

backing for
socket outlet

121 Location of supports

122 Fixing soleplate to a wall

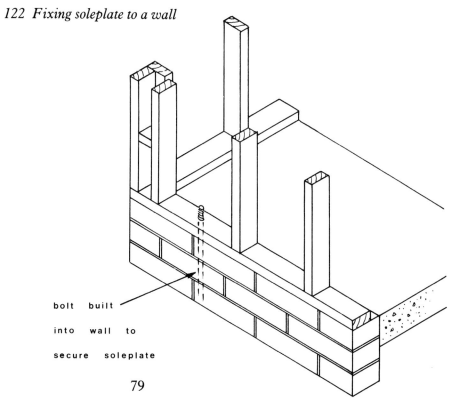

bolt built
into wall to
secure soleplate

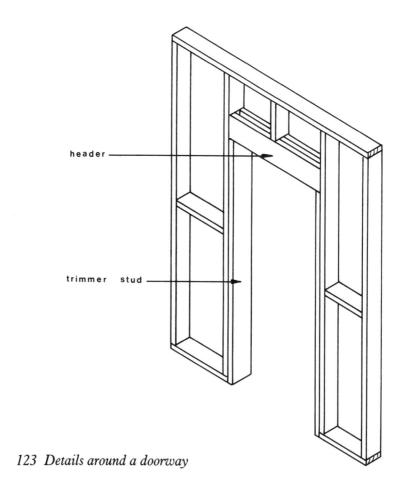

header

trimmer stud

123 Details around a doorway

10 STAIRS

The building of a flight of stairs should present no difficulty for the average home carpenter, provided he is prepared to take time over the various stages of construction. It is essential that a detailed plan is drawn, together with sectional details, and that this is discussed with the local authority to ensure that it will conform with the relevant regulations.

If, for example, it is intended to open up the attic of an existing two-storey house, then there are specific requirements regarding landing size, fire doors etc., and it is clearly necessary to comply with these regulations to preserve the safety of both building and occupants.

Definitions of terms

In order to understand the regulations, it is necessary to be familiar with the various technical terms used. The word *flight* describes the series of steps between floors or between floor and landing (fig. 124). There should be a minium of three steps and a maximum of sixteen for each flight.

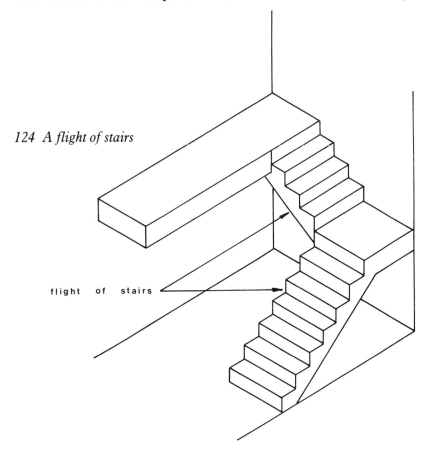

124 A flight of stairs

flight of stairs

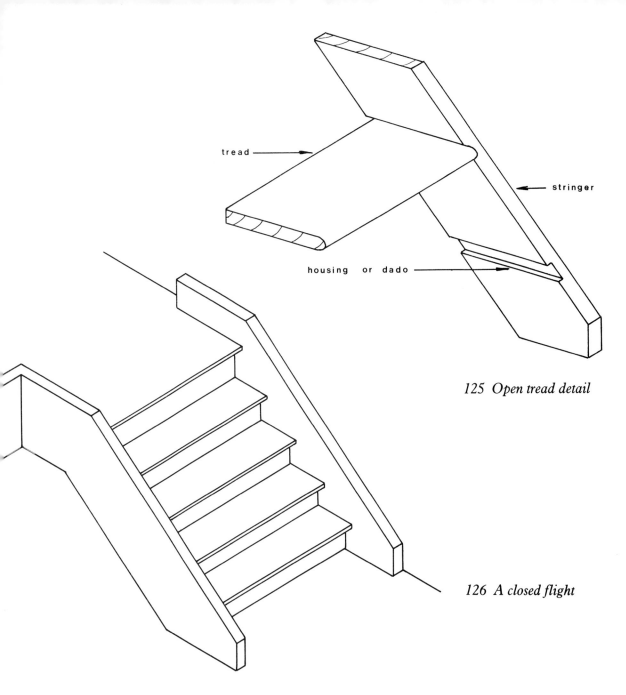

tread

stringer

housing or dado

125 Open tread detail

126 A closed flight

The *steps* may be simply a series of horizontal open treads with a space between, or enclosed steps with a vertical face between the treads, known as the *riser* (figs 125, 126). The horizontal step is referred to as the *tread* with the term *rise* describing the vertical distance measured between the upper surface of two treads. The term *going* is the horizontal distance measured between the nosing of two consecutive treads (fig. 127). The *pitch* is the angle formed between a line joining the front edge of each tread (the *nosings*) and the floor. The preferred angle of pitch is 30° to 35° with a maximum of 42°.

82

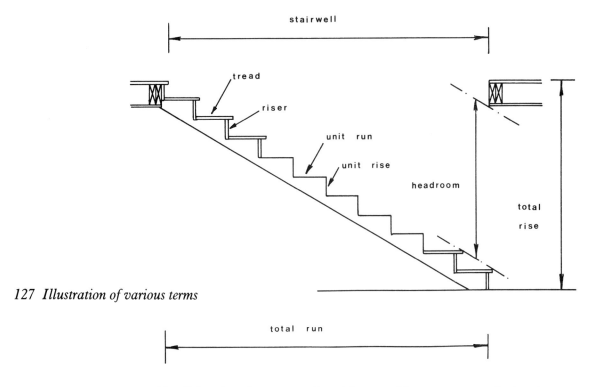

127 Illustration of various terms

Traditional stairs are made with two side supports, called *strings* or *stringers*, that contain and support the treads and risers of a flight of steps.

An alternative is to use stair *carriages*, or cut strings, generally a set of three, which provide a solid shape onto which treads and risers can be fitted (fig. 128).

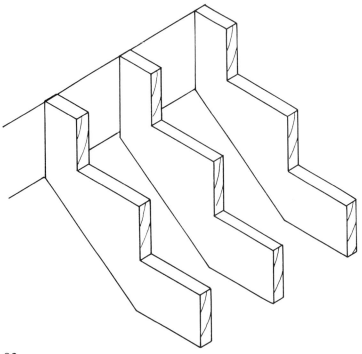

128 Basic carriages

129 A straight flight

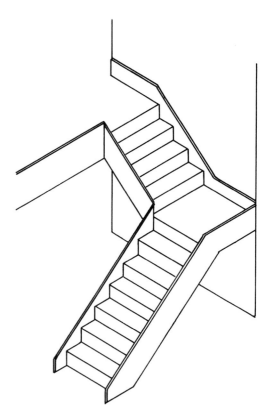

130 A quarter turn

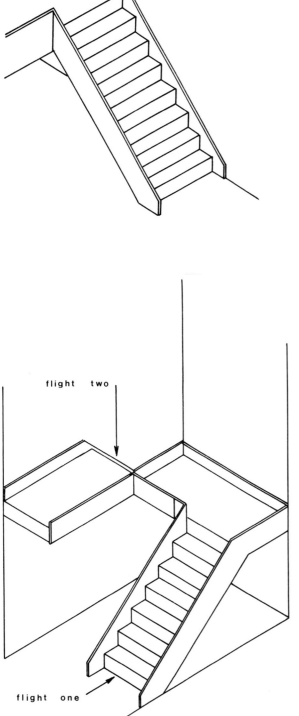

flight two

flight one

131 A half turn

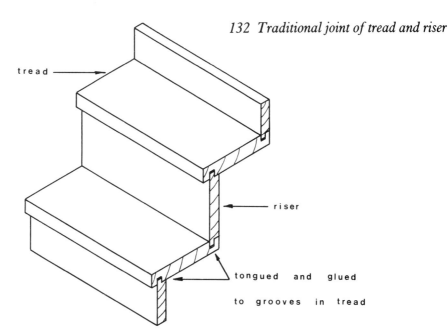

tread

riser

tongued   and   glued

to   grooves   in   tread

Types of stairs

There are three basic ways in which stairs with parallel treads are constructed (figs. 129-130), these being:

    (a) a straight flight
    (b) a quarter or L turn
    (c) a half or U turn

In a traditional English flight of stairs, the riser is tongued on top and bottom, and is glued into grooves cut in the threads (fig. 132). The risers and treads are housed in tapering grooves cut in the face of the strings, where they are glued and wedged.

    An alternative, popular in America, is for the tread to be given some additional support by a groove cut in the riser to house the back of the tread (fig. 133). In either method, the addition of triangular softwood blocks will add to the general rigidity of the flight. The main advantages of the closed-step type of construction are that it is strong and dust-tight, and will not generally develop squeaks.

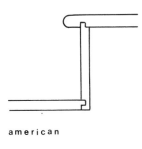

american

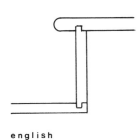

english

133 Comparison of American and English methods

Design

In the design of a new flight of stairs, the first task is to measure accurately the distance between the two levels that are to be connected. Then, estimate the proposed number of risers, and divide the total rise available by this number. In a given flight, it is essential that all the risers are same height, and that all the treads are the same width. There is always one

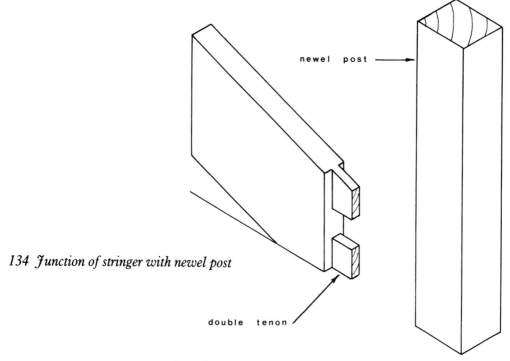

134 Junction of stringer with newel post

newel post

double tenon

more riser than tread, because the landing acts as the top tread.

It is generally accepted that the riser/tread relationship should be governed by the following rules:

    (a) The sum of two risers and one tread should equal 635mm (25in).

    (b) The sum of one riser and one tread should be between 430mm and 460mm (17in to 18in).

    (c) The product obtained by multiplying the height of the riser by the width of the tread should be around 1900mm (75in).

The next stage is to draw a side elevation of the proposed stairs, and to establish the length of the run required, checking that adequate headroom is available. Having completed the calculations for the sizes of risers and treads, the details of construction can now be added.

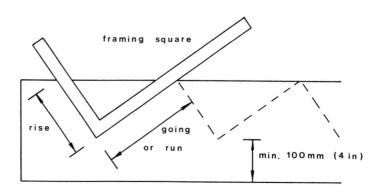

framing square

rise

going or run

min. 100mm (4 in)

135 Use of the framing square

Although it is possible to work from these drawings alone, it is useful to draw out part of the staircase full size. This will help clarify the construction detail, and will be useful in checking the setting-out. Consideration can now be given to the balustrade; that is, the newel post, handrail and balusters. The strings may be fixed to the newel post in several ways: the one most suited to the location should be chosen (fig. 134).

Construction

Very careful marking out will be required, particularly with regard to the strings. For this, a carpenter's framing (or roofing) square will simplify matters – but a suitable template, accurately made, can be an acceptable alternative.

With the framing square, the method is simply to set the square so that one dimension corresponds to the rise, and the other to the tread (fig. 135). A line is then drawn along the outside edge of the blade; one for the riser, the other for the tread. The square is then moved to the next position, and the process repeated until the required number of risers and treads have been drawn.

As the stair begins with a riser, it is necessary to extend the last tread and riser lines to the back edge, and similarly extend the lines at the top (fig. 136). Finally, when the bottom tread is installed, the height of the last riser will be shortened by this amount; to compensate for this, the bottom length of riser must be reduced by an amount equal to the tread thickness.

If the strings are to be the carriage type, then another one or two can be marked in a similar way, and the waste material cut away (fig. 137).

The simplest method of supporting the treads is by attaching supports, or cleats, on to which the treads can rest. Another method, suitable for open treads, consists of simply cutting grooves, or dados, into which the tread will fit, the depth of the groove being kept to one third of the thickness of the string. The traditional construction consists of the string having tapered grooves into which the treads and risers fit. Wedges are then glued and driven into the grooves under the treads and behind the risers (fig. 138). The treads and risers are joined together with tongued and grooved joints, and glued triangular blocks added to ensure rigidity.

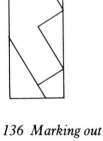

136 Marking out

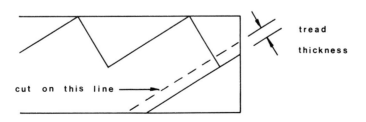

137 Horizontal cut at base of stringer

The marking-out of the tapered grooves is greatly simplified with the use of a template. This can be cut out of hardboard, and should be 1.5mm (1⁄16in) less all round than the ends of the tread, and 1mm (1⁄24in) less all round than the ends of the riser. This allows for cleaning-up and sanding before assembly. It is essential that a set of trial grooves are cut in a scrap piece of timber to ensure that the template is accurate.

Timber for the strings should be well-seasoned, straight grained and, if possible, should be stored for several weeks in the area where it will ultimately be used. Care must be taken in the cutting out of the grooves; the use of a power router will greatly assist in obtaining both straight edges and a consistent depth. The treads and risers should all be cut to the same length, and the appropriate grooves and tongues cut in each, remembering that the bottom riser will be one tread thickness narrower than the remainder.

If one, or both, of the strings are to be mortised into the newel, this should be done at this stage. An initial dry assembly should be carried out to ensure that everything fits snugly into place, and a check made that it will be possible to carry the assembled stairs into the final location. It may be necessary to carry out the assembly in the room where the stairs are to be installed if there is any doubt about taking the assembled unit through passageways etc. The stairs may then be cramped together and any necessary holes for screws drilled through.

For the final assembly, it is easiest to work first on one side only, glueing and assembling with the string upside down and glueing in the top and bottom treads and risers, and ensuring that the assembly is square. This should then be carefully clamped, the wedges glued and driven into place.

When all is firm, the remaining treads and risers may be slid into place, glued and wedged. When the run is complete, the triangular blocks may be glued and pinned into position. When finished, the whole unit should be left clamped for at least twelve hours to ensure a firm and rigid structure.

138 Traditional use of tapered grooves and wedges

timber wedges

taper 1 to 15

CUPBOARDS

The main function of kitchen cabinetry is to provide storage and general convenience while working in the kitchen. Cabinets should be carefully planned and well-constructed so they can cater for varied needs, providing a solid work surface as well as practical and useful storage.

Types of construction

Kitchen cabinets can be constructed *in situ* or can be purchased in kit form, the former generally being built as a frame construction, the latter usually made from pre-cut sheets of laminated board (figs 139-140).

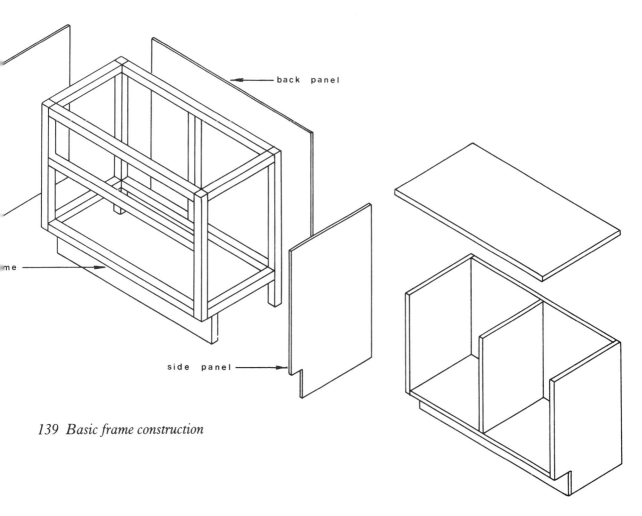

139 Basic frame construction

140 Typical box construction

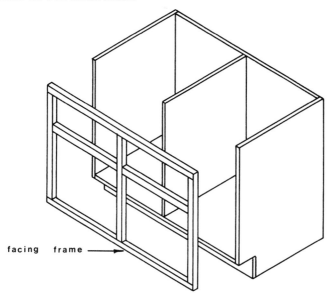

facing   frame ——————▶

The appearance of kitchen cabinets depends almost entirely on the doors and fronts of drawers, as the major part of the construction is not visible. It is well worth spending a little extra time and money on these two areas, and to coordinate their appearance throughout the kitchen.

It is perfectly acceptable to combine the two types of construction so that the finished cupboard looks as though it is of a framed type, which, at the front, it is (fig. 141).

The basic shapes for the non-frame or box construction can easily be executed provided care is taken in the design and marking out. Cabinets can be made piece by piece as built-in or separate units, and then built-in the same way as factory units would be. The choice will depend very much on the individual kitchen, and whether it is being completely remodelled or just a minor addition made.

If a solid wood frame is to be used, then a timber size of 25 × 38mm (1 × 1½in) would be appropriate. This is then covered by thin plywood or hardboard. The box type of construction relies upon the use of 12-19mm (½ to ¾in) material for the main supports, the only solid wood being the drawer guides and runners, and perhaps the face framing.

All base cabinets should be built with a kicker, or toe strip, to take knocks of scuff marks; this generally being provided for by the sides having a notched recess to take the strip. It is also normal for the counter top to hang over the front edge of the unit.

Kitchen layouts

If a new layout is being decided upon for the kitchen, then an accurate plan is essential in order that the various cupboards, units and appliances will fit in. There are three basic layouts for a kitchen (figs 142-144):

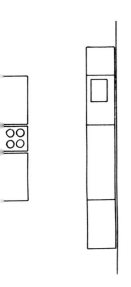

(a) the parallel wall, galley or corridor layout, frequently found in the smaller home when an eating area is not included

(b) the L-shaped kitchen where units are arranged along two adjacent walls, thus leaving an area free for dining

(c) the U-shaped kitchen where a practical and efficient work area is created.

In designing the kitchen layout, some thought should be given to trying to reduce the distance between the 'work centres'; therefore tools, utensils and materials should be as close as possible to the place of use. For example, washing-up liquid should be stored in a wall cupboard near the sink, while seasoning and herbs should be on a rack near the cooker.

Having decided upon the appropriate layout, then the design of an individual unit can be considered. Although most units are of a standard size, they are based upon an average person's height. If possible, it is sensible to try and arrange the height to suit the cook, or cooks! It is generally agreed that the height to the working surface should be 950 to 975mm (3ft 1½in to 3ft 2½in) and that the level of the sink work top should be 50mm (2in) higher.

142 Galley, or corridor, layout

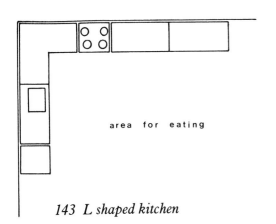

143 L shaped kitchen

area for eating

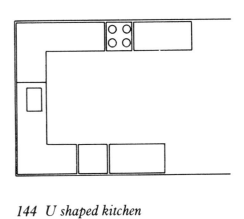

144 U shaped kitchen

Design details

The design of units is varied, but a good base can be made with the box type of construction, and then individual variations added to suit the required appearance (fig. 146). There are four basic kinds of cabinets used in a kitchen, the base unit, wall unit, tall unit and special purpose unit.

A good base unit must be sturdy, with no sticking doors or drawers, and should be designed with a toe space of 75 × 60mm (3in × 2½in). A suitable material for the main box would be 9mm (¾in) veneered chipboard. The increased thickness will enable a saving to be made on the framing.

91

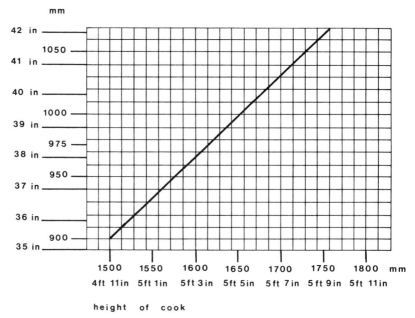

ideal sink height (work-tops 75mm – 3 in lower)

height of cook

145 *Diagram of recommended heights. Worktops outside these minimum and maximum heights tend to be impractical*

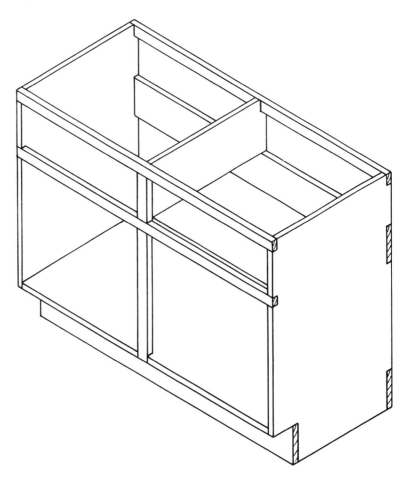

146 *Typical base unit*

92

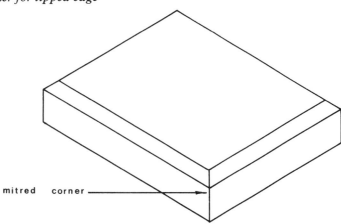

mitred corner

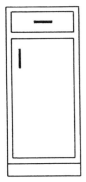

basic

Chipboard with a wood veneer is easy to cut with a fine-toothed panel saw, but plastic laminated chipboard is cut more effectively with a very fine-toothed blade in a power saw. In either case, the exposed edge should be covered, either with a wooden strip, or, if the edge is to be virtually out of sight, a strip of matching laminate (fig. 147). The parts may be joined together with butt joints that are either pinned and glued, or joined with screws and angle brackets. There are several proprietary fittings available for making square joints, but care must be taken where they are used to ensure that they do not interfere with the drawers.

The choice of hinge will depend upon the style of door, whether it is to be inset, flush, or outside the main frame, and whether or not the hinges are to be visible. The type of handle, to both door and drawers, requires further thought; for example, if using a plastic laminate, then a length of aluminium section fitted to the top may be appropriate, but if the finished door appearance is panelled oak, then a timber or black iron fitting would be better.

A range of styles can be obtained by adding various mouldings as a supplement to the hardware (fig. 148). Worktops should be made so that they cover several units, and provide a continuous work surface. Standard sections are available in plastic laminated chipboard, and these are probably the cheapest and most practical solution. However, an interesting surface can be obtained by glueing together smaller timber sections, and then finishing with three or four coats of hard-wearing varnish (fig. 149).

The timber can span in either direction (fig. 150), but it is easier, from a construction point of view, if the sections run parallel to the rear wall. A thicker and deeper section should be used for the front edge, and an interesting finish can be obtained if a contrasting colour is used and the edge shaped with a router.

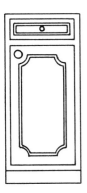

traditional

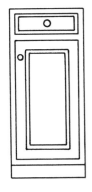

colonial

148 A range of moulding styles

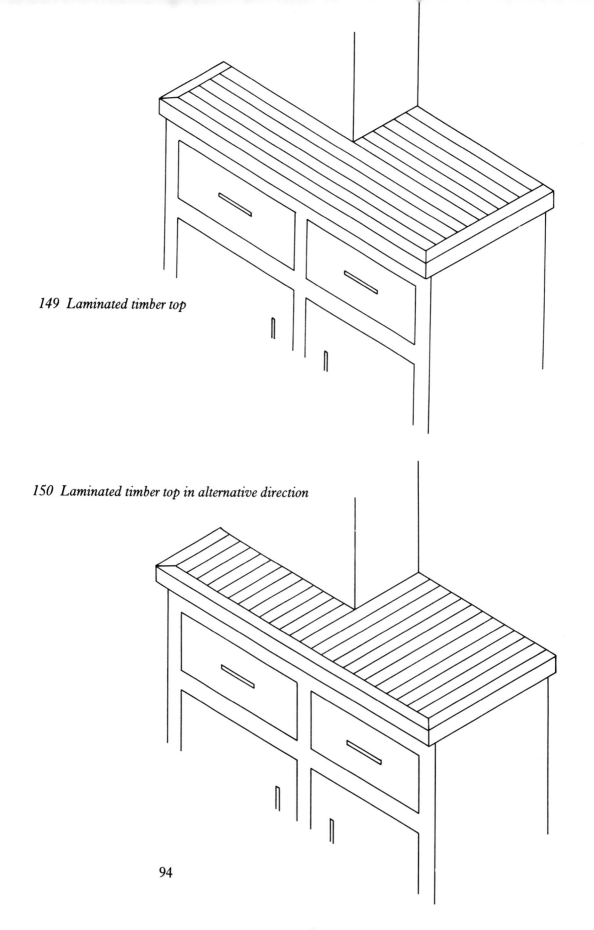

149 Laminated timber top

150 Laminated timber top in alternative direction

Drawer construction

The drawer construction will depend on the general finish of the units, but a standard form of joint will enable drawers to be constructed easily. It is far easier to make four drawers at the same time than individually. Making several at once allows the use of jigs to cut all the sides at once. It is essential that the front and back of the drawer are the same size, thus ensuring that the sides will be constructed parallel.

The runners on which the drawer slides must also be parallel as well as straight and smooth on the sliding surface. The type of runner to be used will influence the construction of the drawer, so one of the following types needs to be chosen.

(a) Rail or batten (fig. 151). This is the traditional method. The drawer is supported by, and slides on, rails underneath the drawer sides, with a further rail at the top to stop the drawer tipping when open.

(b) The three cleat method (fig. 152). This requires less skill and accuracy for construction, but takes up space within the unit. A wooden strip is fitted to the side of the drawer, and two parallel cleats are attached to the frame so the cleat on the drawer can run between. This method is not recommended if the drawers are heavy or likely to carry heavy loads.

(b) Cleat and Groove (fig. 153). This is less wasteful of space and although requiring skill to construct is the normal method used in modern construction. A groove is cut in the side of the drawer, and a single cleat fitted to the frame along which the groove can slide. The groove normally stops short of the front of the drawer.

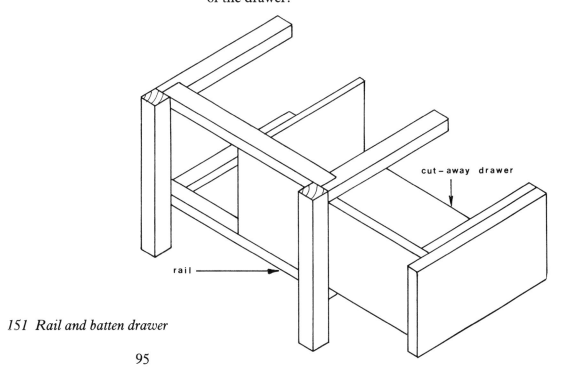

cut – away drawer

rail

151 Rail and batten drawer

95

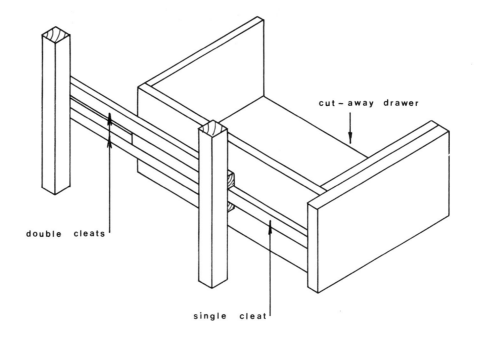

152 Three-cleat drawer support

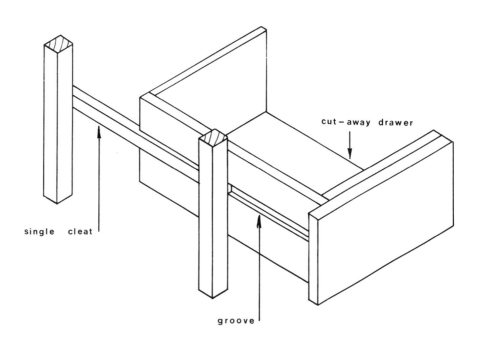

153 Cleat and groove drawer support

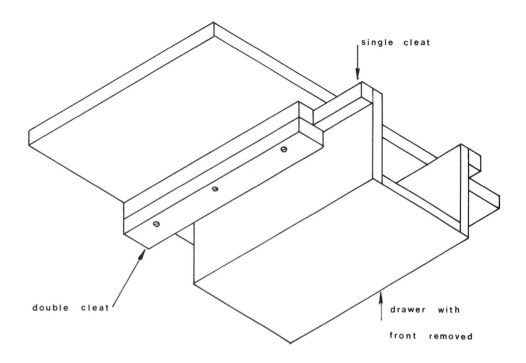

single cleat

double cleat

drawer with

front removed

154 Top-hung drawer support

(d) Top hung (fig. 154). Usually used when a drawer is to be fitted
to the underside of a table-top without any available support
beneath. A single cleat is fixed to the side of the drawer, and a
pair of cleats forming an L section are fixed underneath the
table-top so the single cleat can fit into it. This again is not
suitable for heavy drawers or loads as the fixings tend to pull
out. Heavy drawers should always be properly supported from
underneath.

Most drawers are purely functional, and as long as the drawer is strong
enough and suitable for its purpose, the simpler the design, the better.

As it is only the front that is generally noticed it is essential that this
presents as neat an appearance as possible. For example, do not mount a
flush drawer by the three cleat method, as this will leave large and
unsightly gaps either side.

When working out the size of drawer needed to fit into a given space, it
is advisable to leave 3mm (⅛in) clearance all round to allow for the
changes in humidity and temperature, and subsequent expansion and
contraction of the wood.

12 SHELVES

Shelves have to support a variety of loads, and must be designed so that they are both functional and elegant. The section on design indicated that a little thought before starting construction will avoid a lot of problems. For example, it is essential that shelving is of adequate thickness and strength to withstand the load, and that the span is realistic.

The actual shelf may be of either solid material, slatted timbers, or manmade boards – but it may be necessary to provide some additional thickening to the front edge (figs 155-157). A touch of elegance can be introduced if the front thickening is of a varied section, particularly if a curve can be incorporated.

There are many ways of providing support to the ends of the shelf (figs 158-160). The most commonly used are:

(a) bearers fixed to the wall
(b) brackets, either of metal or timber
(c) timber frame uprights
(d) solid uprights

Of these, (a) and (b) rely upon a fixing to the wall, whereas (c) and (d) may be free-standing.

Fixing

Fixing directly to a wall is often difficult, but there are several ways of overcoming the problem. The most popular is to use some form of plug.

The type of plug used will depend upon three basic needs:

(a) the required strength of fixing
(b) the type, condition and density of the base material
(c) whether the base material is of hollow or solid construction

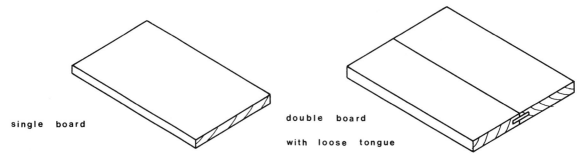

single board

double board
with loose tongue

155 *Solid shelving*

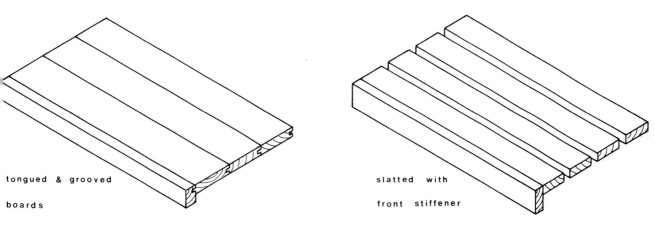

tongued & grooved
boards

slatted with
front stiffener

156 Boarded and slatted shelving

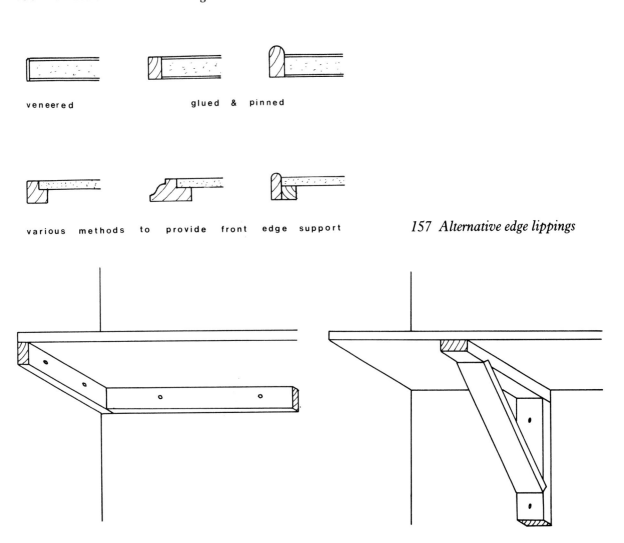

veneered

glued & pinned

various methods to provide front edge support

157 Alternative edge lippings

158 Bearers fixed to the wall

159 Typical timber bracket

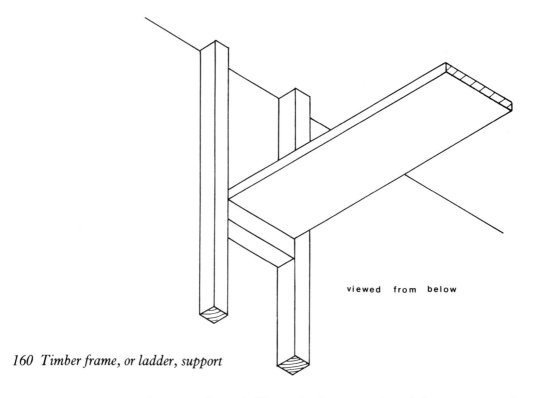

viewed from below

160 Timber frame, or ladder, support

Plugs are of wood, fibre, plastic or metal, and there are several specialist suppliers who provide a wide range suitable for most applications. Most famous is Rawlplug, and it is possible to find a suitable type from the wide range that they produce. It may be possible to make a suitable plug from scrap wood, but care must be taken both in its manufacture and application (fig. 161). Plugs should not be used in situations that may disturb the surrounding brickwork, for example near a corner or towards the top of a wall.

Plastic plugs, either from nylon or polythene, have good holding characteristics and are not affected by changes in temperature or humidity.

If the base material is of a hollow or thin construction, then a special cavity type fixing will be needed for most situations.

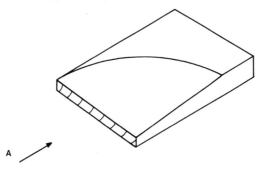

A

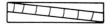

view from A

161 Timber plug

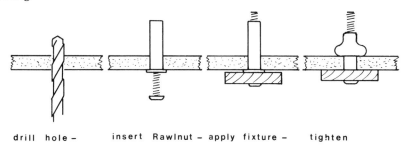

drill hole –   insert Rawlnut – apply fixture –   tighten

A Rawlnut, which has a flanged rubber sleeve housing a nut and bolt, compresses as the screw is tightened, thus forming a flange behind the wall (fig. 162). An alternative is the spring toggle, which consists of a screw attached to steel spring wings, which can be folded back, pushed through the hole in the cavity where they spring apart, and then drawn back against the base material by tightening the screw (fig. 163).

An alternative is the gravity toggle, which has an off-centre metal channel instead of a nut, which, when pushed through the cavity, hangs vertically, thus allowing it to be tightened (fig. 164).

The Rawlanchor is similar to the Rawlnut, but has a plastic plug which compresses as the screw is tightened (fig. 165).

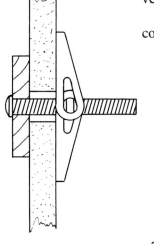

163 *A spring toggle*

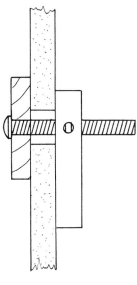

164 *A gravity toggle*

165 *Section of a Rawlanchor plug*

101

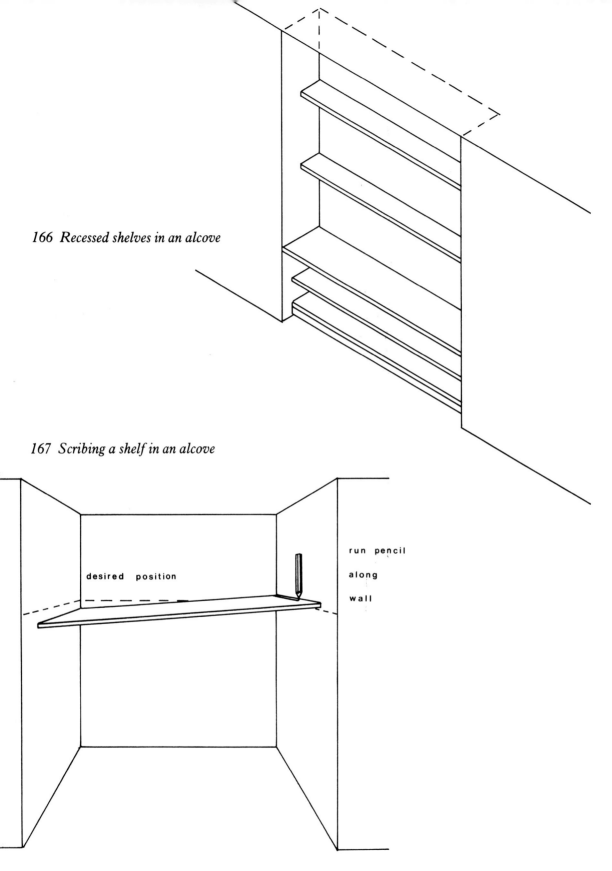

166 Recessed shelves in an alcove

167 Scribing a shelf in an alcove

desired   position

run pencil

along

wall

Alcoves

It is obviously simpler if shelves can span from two solid walls and frequently homes are designed with alcoves that provide suitable support.

A set of shelves can be built so that the front is flush with the wall at either side of the alcove, but, as this creates the effect of a flat wall, it is preferable to set some of the shelves back a little way, as this will retain the effect of the alcove (fig. 166).

One problem often encountered is that the alcove is not square, and it may be necessary to scribe the shelf in order to obtain a fit. To scribe a shelf, it is first necessary to measure the maximum dimensions of length and depth of the alcove, and to cut a piece of timber a little more, say 3mm (⅛in) than the measured dimensions, so the shelf is a fraction too long and too deep.

The end of the board that touches the smoothest and straightest wall should be scribed first, and the minimum amount of timber removed. To scribe, jam the board into the alcove as near as possible to the desired position, and then mark holding the pencil vertically, directly against the wall (fig. 167). This should be repeated at the other end – and, if correctly done, will allow the shelf to be fitted horizontally so that the longer side at the rear can be done in a similar manner.

It may be possible, if the alcove is large, for an individual unit to be constructed and placed within the area, but this will depend upon the need for shelving space, and what it will be used for.

Display

The modern trend is for a set of shelves to be used for general display of souvenirs, ornaments and knick-knacks, and the second-hand shops have been kept busy trying, for example, to locate old print trays that provide a series of small compartments ideal for this sort of display (figs 168, 169). To reproduce this type of unit is very time consuming, but it presents an interesting challenge to the patient worker (fig. 170).

By adopting the principle of the printer's tray, but at a larger scale, the woodworker can achieve a very individual display unit. For people with a specialized display, the unit is secondary to the objects, but for the average home, the unit is as important as the varied objects it contains (figs 171, 172).

Kitchen

In the kitchen, shelving units have to be designed for purely functional purposes, but this still allows the varying shapes to be based upon pleasant proportions and appearance (fig. 173).

This type of unit is easily constructed from solid timber (parana pine is

168 Printer's tray used as display unit   169 Corner detail of printer's tray

170 An alternative layout for display of small items   171 A modern display unit

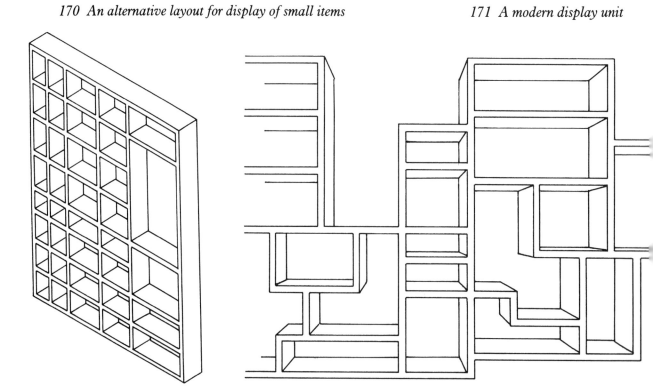

104

172 Modern shelving unit

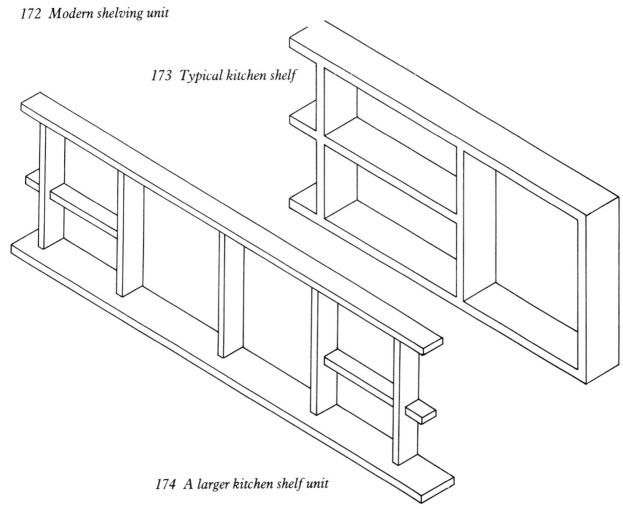

173 Typical kitchen shelf

174 A larger kitchen shelf unit

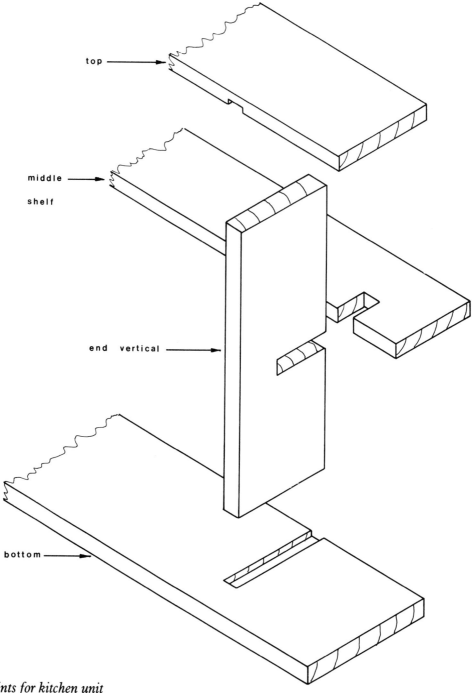

top

middle

shelf

end   vertical

bottom

175 Details of joints for kitchen unit

suggested) as, although the thickness of the timber gives a solid
appearance, this is softened by the light, mellow colour of the wood (figs
174, 175). As an alternative, the same design may be adapted for a
veneered or laminated board, with suitable lipping, either of a matching
colour or as a contrast (fig. 176).

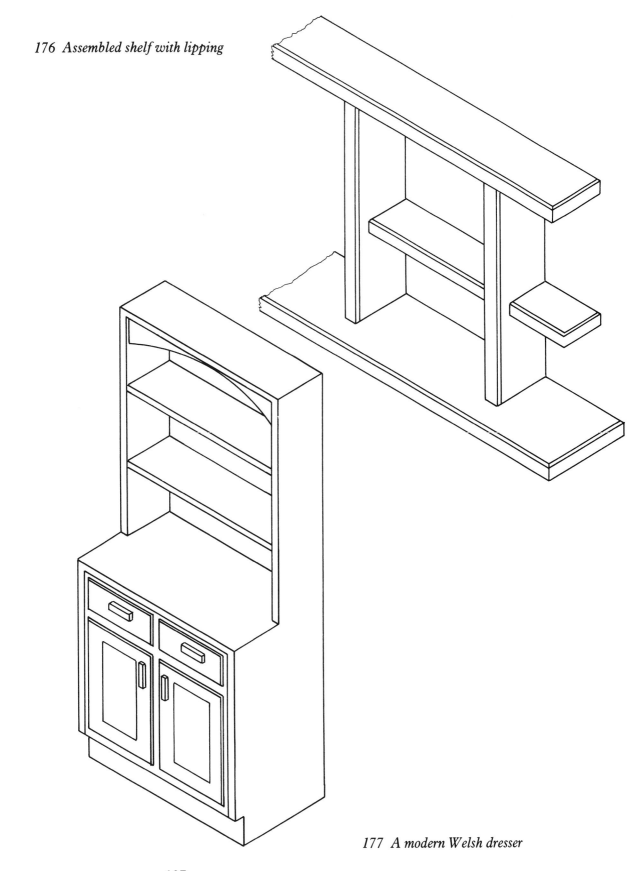

176 Assembled shelf with lipping

177 A modern Welsh dresser

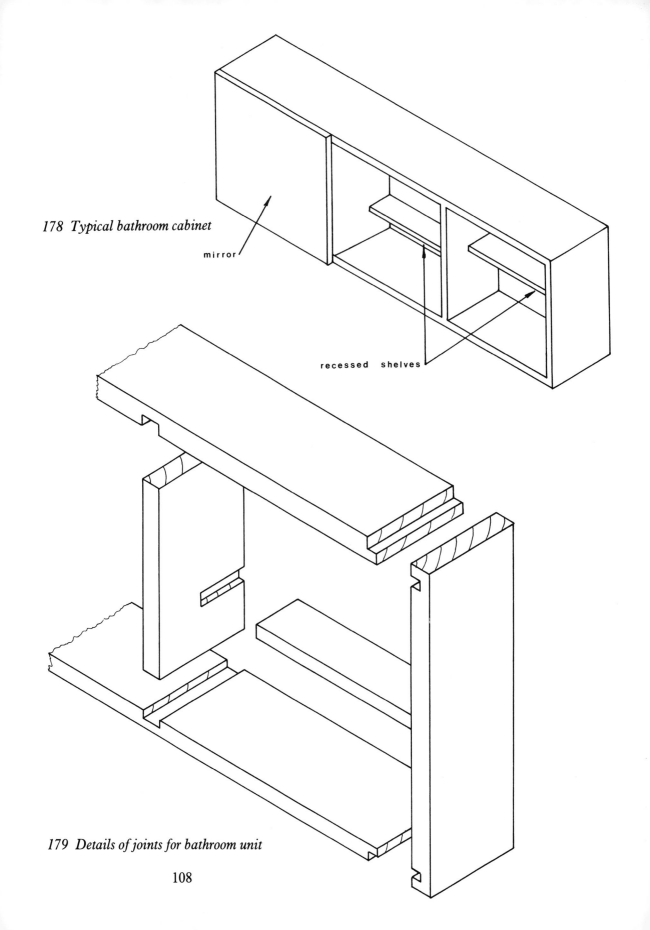

178 *Typical bathroom cabinet*

mirror

recessed   shelves

179 *Details of joints for bathroom unit*

The fitting of this unit to the wall can be easily done by plugging and screwing a batten of a length to correspond with the inside measurement, and allowing the unit to rest on, and be fixed to, this batten. For the sake of appearance, the batten should be of a similar material to the main carcase.

A more elaborate set of shelving, on the style of a Welsh Dresser, can make an attractive feature as well as being very functional (fig. 177). The first step in this construction is to cut the sides square, and to the required length. The stopped housing joints can then be marked to position, the shelves and top cut to suit, and the housing grooves made to match the shelf thickness. The shelves and top can be cut with a rebated groove near the back edge so that plates can be displayed, the groove being 6mm (1/4in) deep and 13mm (1/2in) wide.

Some form of panelling can be added to the rear, and the curved arch fitted to provide the finishing touch to the set of shelves.

Bathroom

A set of shelves combined with a cupboard is a useful addition to the bathroom. It is normal for a mirror to be added to the front of the cupboard door (fig. 178). As condensation is often a problem in the bathroom, it is essential that a waterproof glue is used for the assembly.

Plastic laminated chipboard is a suitable material for the main carcase, and, if the front is lipped with timber, this can be painted in a colour that will either match or blend with the general colour scheme, as well as providing the necessary base for fitting the hinges and catches (fig. 179).

As the loading of this set of shelves will be relatively small, the use of 'mirror' brackets can be considered, although the addition of a small support batten under the bottom of the unit is recommended.

General living areas

Shelving in the lounge, study and family rooms should always be rugged and sturdy (figs 180, 181). Preferably it should be part of a floor to ceiling unit, the spaces being designed and allocated for specific purposes. The cost of using solid timber for this type of unit may be considered excessive, but a compromise of using veneered chipboard or blockboard with a substantial front edge of matching timber is often acceptable. If the edge can be shaped and routered with a step cutter, then the finished product may well be more stable than one of solid timber (fig. 182).

Careful detailing of all joints and junctions will greatly improve the appearance, and can transform a set of shelves into an elegant piece of cabinet making (fig. 183). Although not of the same elegance, a set of storage shelves for the workshop or garden shed should be considered in the same way; that is, the usable areas should be designed for the purpose

180 Simple set of shelves with individual support

181 An attractive double shelf

182 Detail of joint and edge finish

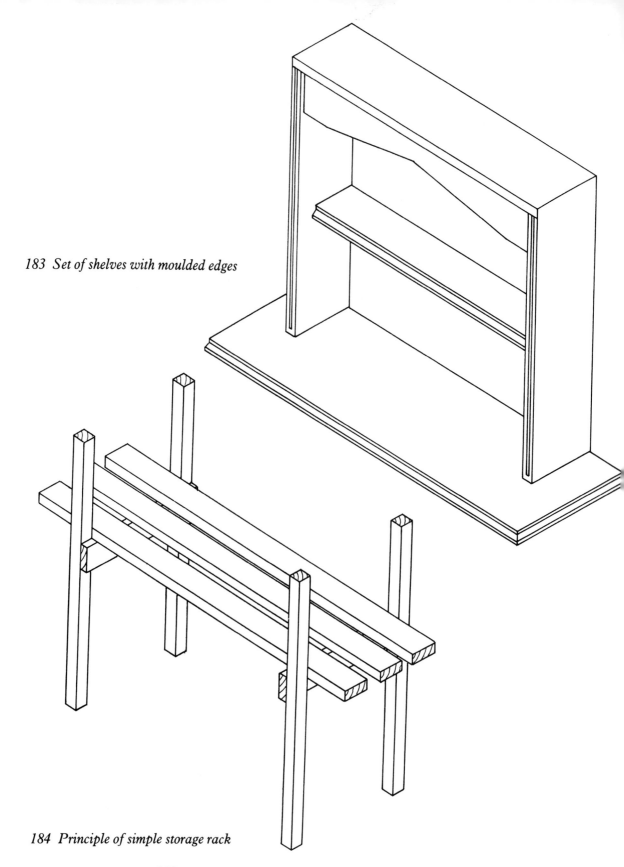

183 *Set of shelves with moulded edges*

184 *Principle of simple storage rack*

185 Ladder storage shelving

intended, and careful attention be given to the construction of joints (fig. 184).

There is no reason why rugged workshop or storeroom shelving should not have a form of elegance, and a little extra time spent in consideration of the design will be well repaid, both from a functional, and an aesthetic point of view (fig. 185).

13 THE WORKSHOP

The allocation of a specific workshop area is often regarded as a luxury, but, in reality, it should be the aim for every handyman. It is highly desirable that tools, equipment and timber should be housed together, and that adequate working space is available. Unfortunately, there is often the tendency to put off setting up the workshop on the grounds that there are far more important maintenance and construction jobs to be done.

Far too often, the handyman is forced into this position. As a result, tools get lost, damaged or broken, and the simplest of tasks takes far too long to accomplish, as time is spent hunting for tools, as well as having to clear away everything whenever there is a break in the work. If a workshop area is carefully planned, and storage of tools and equipment carefully arranged, then a job can be tackled very much more quickly, a start being possible without having to hunt around the whole of the house to find a screwdriver, or whatever is needed (fig. 186).

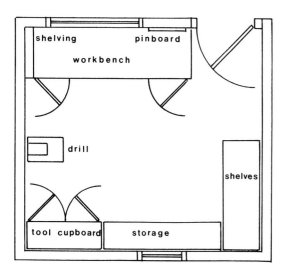

186 Basic layout for workshop

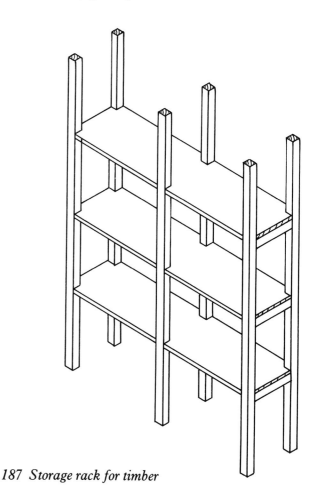

187 Storage rack for timber

The workshop should be light and airy, well ventilated, and with as much free working space as possible. Although it is always nice to have a handy store of off-cuts and other timber, care must be taken to see that these do not become untidy, as the convenience of immediate access is then lost.

The provision of a tea-chest for 'rejected' timber and off-cuts will encourage tidiness, and also allows for the periodic clearing out when a decision can be made for obviously useless pieces to be consigned to the fire. Longer lengths of timber should be carefully racked, so that warping and twisting is avoided as far as possible (fig. 187).

Unless a large amount of storage space is available, the home handyman is advised to restrict the amount of readily available timber to a minimum – although a chance to collect a supply of the more interesting timbers should never be passed up.

The workbench

The workbench is the major item in any workshop – the essential requirements being that it is rigid and solid, and large enough for the type of work to be carried out. Various designs have been in and out of fashion, and each carpenter, joiner or craftsman will have his own individual preference. However, many features are common to all: a solid, flat work-top, a recess for tools, a vice, and some form of bench stop (fig. 188).

The actual working height will vary for individuals, but an average height of 810 (32in) should be normally acceptable.

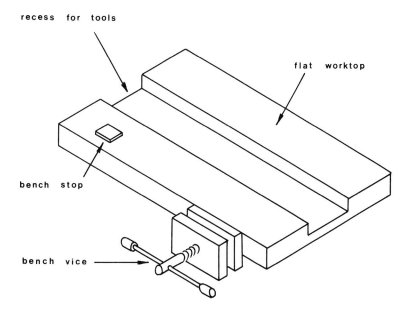

recess for tools

flat worktop

bench stop

bench vice

188 Requirements for workbench top

115

189 *A typical bench vice*

The vice, although an expensive item, is a very necessary piece of equipment. Nowadays they are normally a metal casting, although the traditional form was of wood (fig. 189). Metal vices are sold without a wooden lining to the jaws, so these will need to be made as soon as a vice is purchased, beech being an ideal timber for this job.

The fitting of the vice to the bench will depend on the type of vice purchased, but adequate instructions will be included with any reputable make. The construction of the bench should be related to its future use; for example whether it will be a permanent feature, or whether it will need to be transported to another location at some time. Should it be likely that it will need to be moved, then the design should be such that it can be

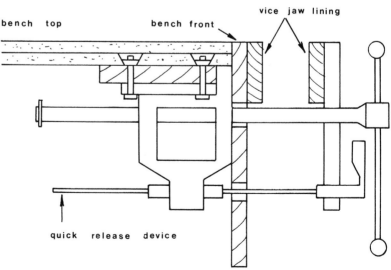

190 *Section through a vice*

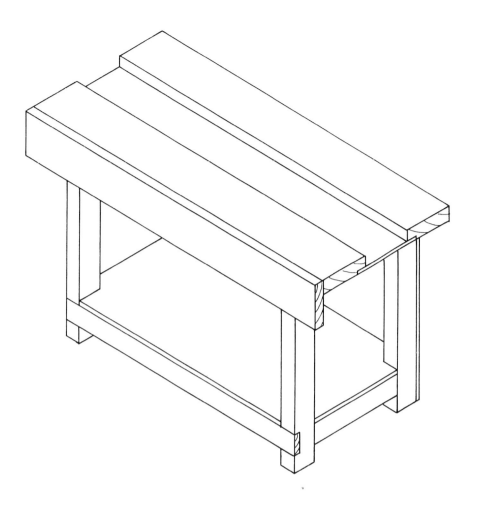

191 A rugged bench

readily dismantled and transported to a new location. Details of the construction of a bench that is both solid and rigid, but can be easily dismantled, are shown (fig. 191).

The basis of construction is the two A-frames, these being held in position by the top working surface. The solid bottom shelf is fixed to supports on all four sides, together with the rear plywood panel (fig. 192). The whole assembly could be glued and screwed to provide a permanent feature of the workshop, but if the bottom front and rear supports, the bottom shelf and the rear panel are screwed only, then these could easily be removed.

The top assembly can be either screwed or bolted to the top horizontal members of the A-frame: in either case the top of the fixing method would be countersunk. This would thus allow the workbench to be dismantled into flat units that could be easily transported and reassembled in a new location. The A-frame is constructed using basic mortise and tenon joints that are glued and pinned to provide the solid basic support to the bench (fig. 193).

117

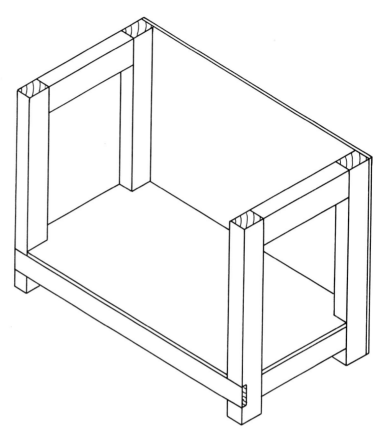

192 The base frame for the bench

Storage of tools

The storage of tools can create problems, but whichever method is chosen it is essential that the tools are returned to the correct location after use. If a large cupboard is available, it can be adapted so that each tool can be allocated a space, either clipped on to a vertical face at the back, side or door, or in a specially-designed rack either free-standing or in a drawer (fig. 194).

An alternative to the cupboard is to use a solid baseboard, fixed vertically to the wall, with individual holdings for each tool, the shape of the tool being drawn on the board so that its location is easily found after being used (figs 195-196).

Before deciding on the best type of storage for a particular workshop, the first priority is to determine the tools to be stored, and any possible future items. If these are all laid out, they will present a formidable display, but, by grouping items together for racking – those requiring individual attention, and those that may be stored together – some form of order can be obtained, and the most practical use of the space available can be found.

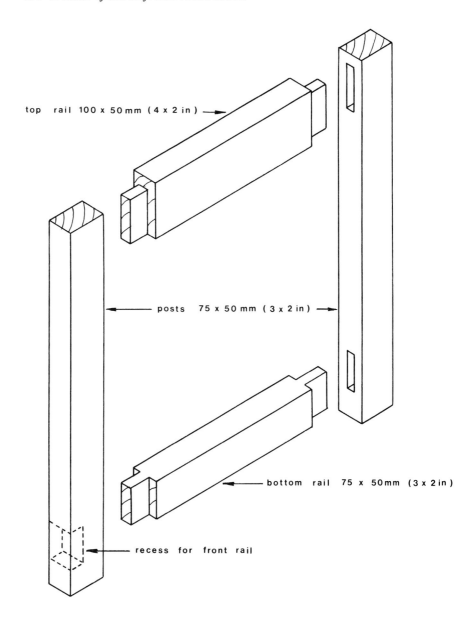

top   rail 100 x 50mm (4 x 2 in)

posts   75 x 50 mm (3 x 2 in)

bottom   rail   75 x 50mm (3 x 2 in)

recess   for   front rail

The construction of some form of tool box may well help, in that a multitude of small items can be allocated a space within a relatively small volume, particularly if trays can be incorporated. One problem with trays is that the tool box must always be kept upright. This is sometimes difficult if the tool box is moved around to various locations. The construction of a combined tool box and working support will help with this problem, as it is very clear which way up the unit should stand (fig. 197).

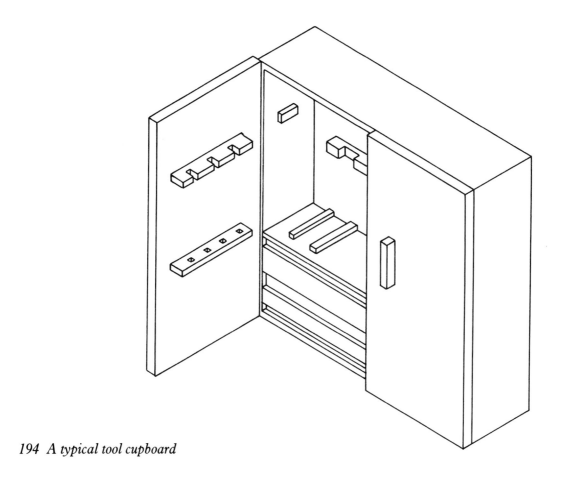

194 *A typical tool cupboard*

draw   the   outline

for   each   item   and

fix   appropriate

clip   or   support

195 *Baseboard for storage of tools*

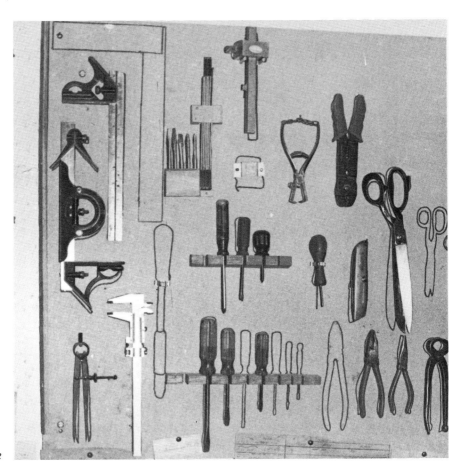

196 *The baseboard in use*

197 *Combined storage and work platform*

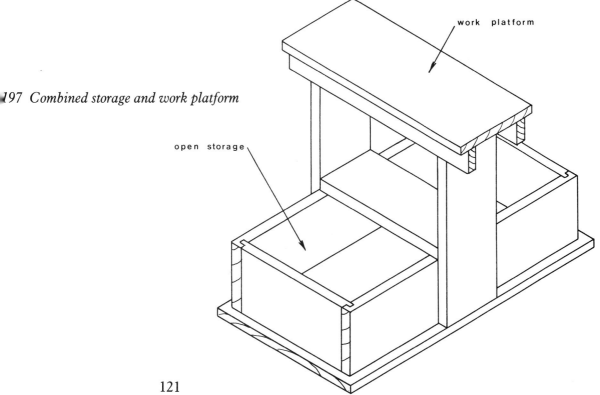

work platform

open storage

121

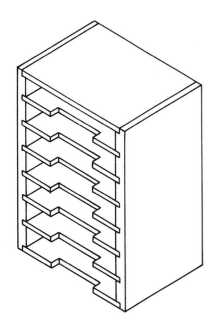

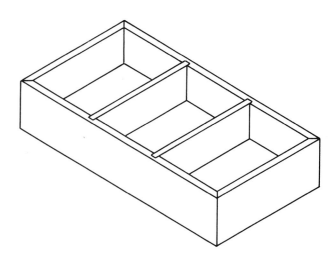

198 Storage rack for glasspaper　　　　　*199 Tray for storage of used glasspaper*

Storage of the ancillary equipment needed in a workshop can present further problems; in particular the large collection of tins containing paints, varnish, stains etc. It is very tempting to keep the last few drops in a tin, but, after a relatively short period, a skin can form and the quality deteriorate. One answer is to empty the left-overs out of the tin and into a glass jar, label the jar with a note of what it contains and when it was put in, and then store on an open shelf.

This will have the dual benefit of both reminding what is available, and providing a visual check on the contents. These jars can be kept on shelves out of the way, thus utilizing wall space to good advantage. Another item that needs to be readily accessible is glasspaper. As it comes in a wide variety of standards and grades, it is particularly helpful if it can be stored in a set of shelves that allow for the separation of the various types (fig. 198).

Used pieces of glasspaper are often needed, and a small box with perhaps three compartments will be found to be particularly useful. The grades can be kept as rough, medium and fine, and each piece put into the appropriate compartment (fig. 199).

Reference material is always needed in a workshop, whether it be a plan of the current project, or a book that contains well tried recipes for staining, varnishing or other finishes, and these should be considered when planning the design. A set of shelves is easily constructed to allow for the occasional and permanent storage of books and material, and, although liable to collect dust, is a useful addition (fig. 200).

122

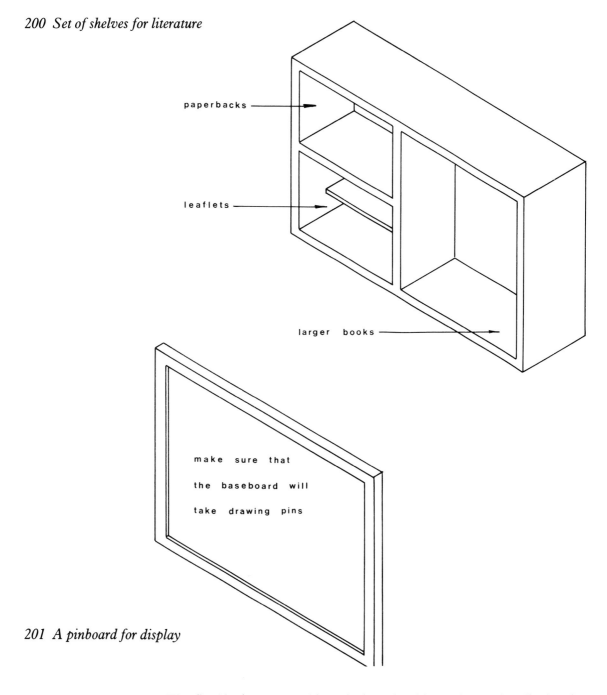

paperbacks

leaflets

larger books

make sure that

the baseboard will

take drawing pins

201 A pinboard for display

The final item suggested is a pin-board. This can be used to display the details of any project, as well as being used as an *aide mémoire* for the odd idea or need that crops up (fig. 201). For example, if one was running low on say, 25mm (1in) panel pins, a little note pinned to the board will jog the memory next time a visit is to be made to the local hardware centre.

The board can also be used to display some of the more common conversions from metric to imperial, or for the display of information concerning stock sizes of material, prices, suppliers' phone numbers etc.

14 GARDEN SHED

Before embarking on the construction of a garden shed it is important to decide exactly what it is to be used for. If it is to be solely for storage of garden equipment, then its location must be such that it is convenient to use, but not obtrusive. If it is to be used as a potting shed, then the location and orientation will need to be considered, as well as easy access and availability of water. The size will be governed by the required use (fig. 202), as well as the amount of money available – but it should be designed as large as possible.

It is amazing how much will be collected within a short time, and, unless space has been well thought out and planned, the resulting chaos will have the opposite effect to that intended – instead of items being readily available, they will rapidly be lost under a growing pile of junk. The ambitious handyman may well want to experiment with new shapes and techniques for his garden shed; after all, it is one of the few items that appear to be exempt from building bye-laws and planning rules. Round or square, pitched or flat roofed, clapboard or prefabricated modules; the scope is endless and could well form an interesting design project for the winter evenings (figs 203, 204).

202 The size must be considered

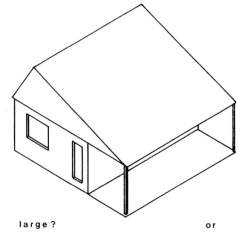 

large ?　　　　　or　　　　　small ?

203 The shape must be attractive

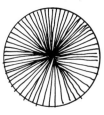 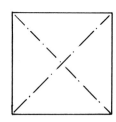

round ?　　　　or　　　　square ?

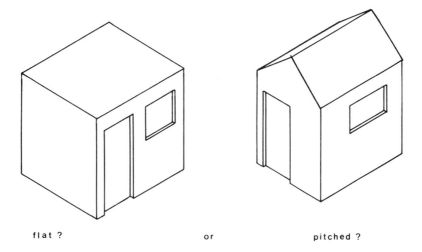

flat ?   or   pitched ?

Basic details

However, it is assumed that a relatively traditional approach is being considered, and that the fundamental requirement is to house garden tools, bird and shade netting, posts etc., and also to have a small bench available for potting, propagating and seeding boxes (fig. 205). Probably the most important item will be the garden mower, whether electric, petrol or diesel, and this should have a place of its own, together with space for the accessories, lubricants, fuels etc.

One safety feature of the shed detailed is that the main inflammable items are stored separately, and that the only access to them is from outside. Another detail included is the provision of a porch by the door.

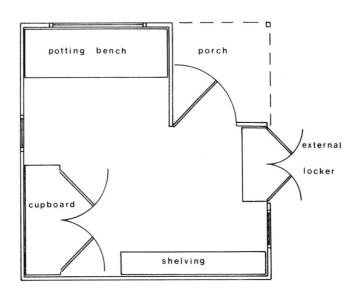

205 The basic layout

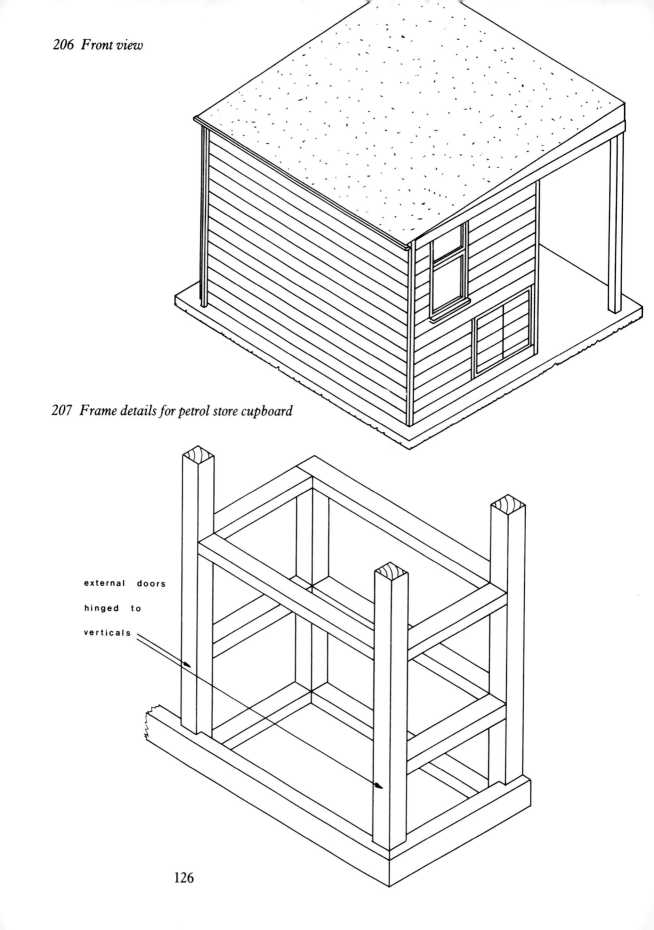

206 Front view

207 Frame details for petrol store cupboard

external doors

hinged to

verticals

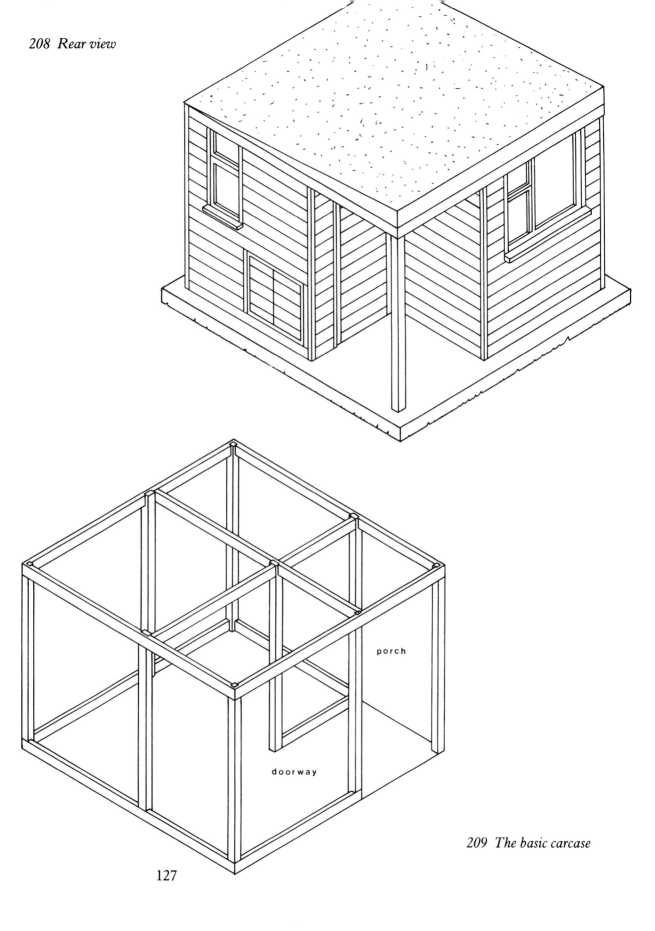

208 Rear view

porch

doorway

209 The basic carcase

127

This has been done to provide the occassional refuge when caught by a sudden shower, as well as allowing comfortable entry and exit into the shed at all times (figs 206-208).

The carcase of the shed is based upon post and beam construction, with standard joints to the four main corner posts, and the intermediate supports being cut in a similar manner (fig. 209). It is assumed that a sound flat concrete base can be provided, and that the floor joists will be supported on a brick, with a slate damp-proof course set between the brick and timber (fig. 210).

Diagonal supports are provided to ensure the necessary rigidity to the structure. The walls are covered with feather-edged boarding, the roof being of plywood and covered with three layers of bituminous roofing felt, the top layer being mineralized, with a thin coating of stone chippings.

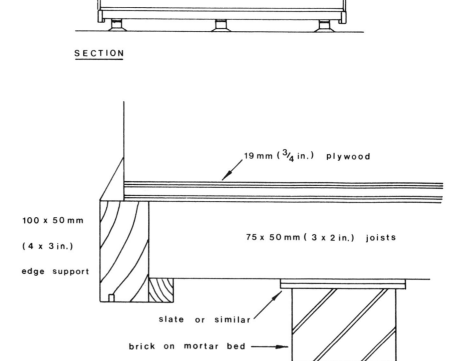

210 Details of floor/wall junction

In order to allow the maximum use of the shed throughout the year, consideration should be given at the design stage to some form of insulation. A fibreglass quilt, 50mm (2in) in thickness, could be included in the wall construction, and the interior lined with plasterboard or hardboard, and a similar construction included beneath the plywood roof. An alternative could be to omit the inner lining and to carefully cut out sheets of polystyrene to fit between the main timber members.

Construction

The first step is to decide upon the required dimensions, and to draw out the separate elevations with these dimensions included, together with the sizes of the timber members (fig. 211). From this basic drawing, the details of the individual joints can be decided upon; wherever possible, keeping to a standard type so that several members can be marked out and cut at the same time.

From the elevation shown, posts A, B and C will be of equal length, and will have similar joints cut at the top and bottom (fig. 212). The horizontal member, both at the top and bottom, will be positioned so that half its section is housed within the post (fig. 213). A and B will have exactly the same marking out, top and bottom, whereas post C will have similar marking to A at the top, but nothing at the bottom (figs 214, 215). In order to avoid some of the problems of rot, a simple base support has been included, that also allows for some flexibility in installation.

211 Typical elevation detail

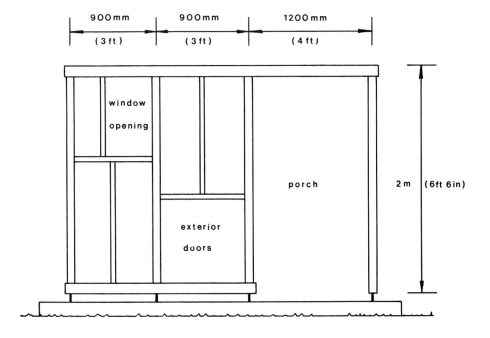

129

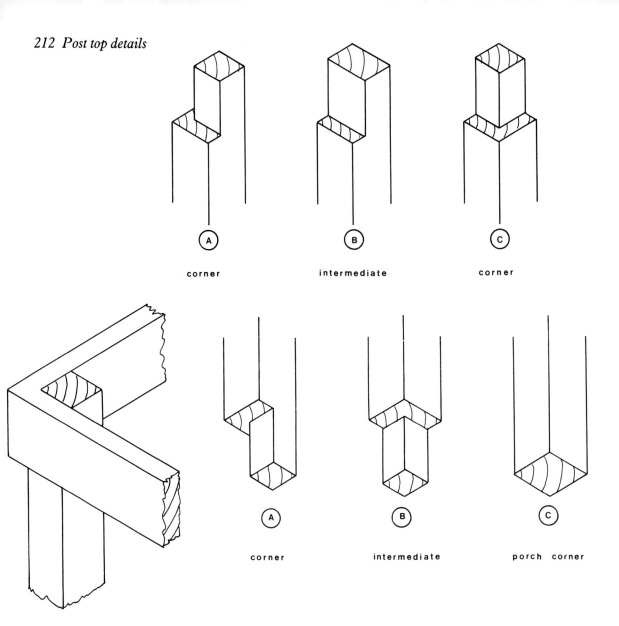

212 Post top details

A
corner

B
intermediate

C
corner

A
corner

B
intermediate

C
porch corner

213 Typical top corner support

214 Post bottom details

Pockets may be left in the base slab when it is cast; a simple 50mm (2in) cube of timber or polystyrene, set in the correct position, will suffice, or a hole may be cut or drilled after the slab has been cast. At the base of each post that will have support, a hole, large enough to take the thread of the bolt, should be drilled some 75mm (3in) into the wood. This will allow for any final adjustment to the length of the bolt beneath the main support plate. This square plate, with the centre hole threaded to match the coach bolt, is fixed to the bottom of the post with four, coarse threaded screws, 60 to 75mm (2½ to 3in), set in pre-drilled holes.

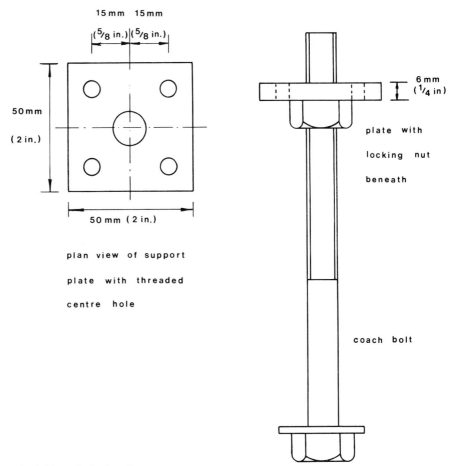

15 mm   15 mm

($^5/_8$ in.)($^5/_8$ in.)

50 mm

(2 in.)

50 mm (2 in.)

plan view of support

plate with threaded

centre hole

6 mm
($^1/_4$ in)

plate with

locking nut

beneath

coach bolt

215 Base bolt detail

The coach bolt can then be attached to penetrate some 50mm (2in), with the nut positioned below the plate. When the posts are held in position, the bolt can be adjusted up or down until the posts are all at the same level. The nut can then be tightened up to the underside of the plate and the head of the bolt grouted into the concrete base.

The horizontal members should be carefully screwed to the main posts, and the intermediate members on each elevation added. Care must be taken to ensure that they are firmly attached, and a check kept to see that they are either vertically or horizontally correct (fig. 216). It is at this time that diagonal bracing may be added. The design of this particular shed is such that only three main diagonal braces will be needed if the joints have been accurately made (fig. 217). However, it will be easy to see if the structure is stable and if any additional bracing is needed (fig. 218). The main roof joists can be cut to length and fixed, each one being strapped down to the main horizontal member, the firring pieces added, and the plywood sheeting securely nailed on (fig. 219). The roofing felt should be laid in accordance with the manufacturers' instructions; particular care

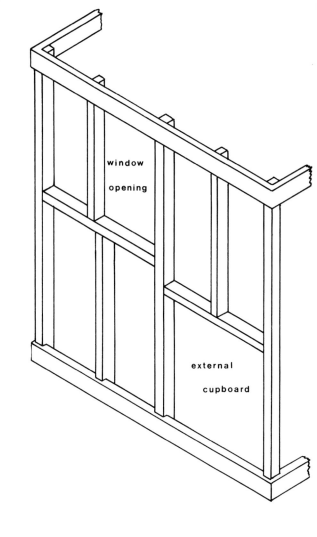

diagonal

braces

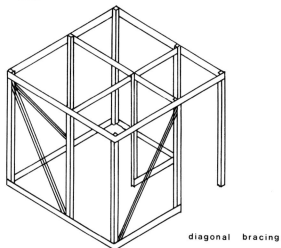

diagonal   bracing

217 Location for diagonal bracing

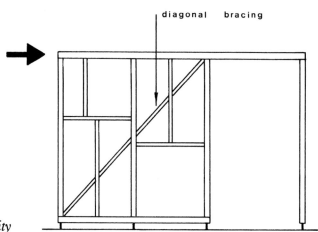

218 Testing for rigidity

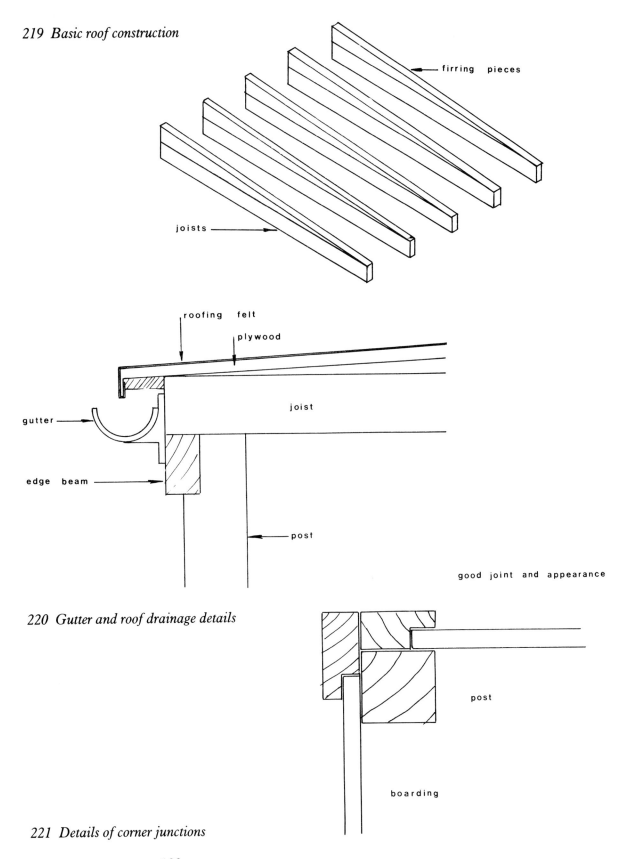

219 Basic roof construction

firring pieces

joists

roofing felt

plywood

gutter

joist

edge beam

post

good joint and appearance

220 Gutter and roof drainage details

post

boarding

221 Details of corner junctions

133

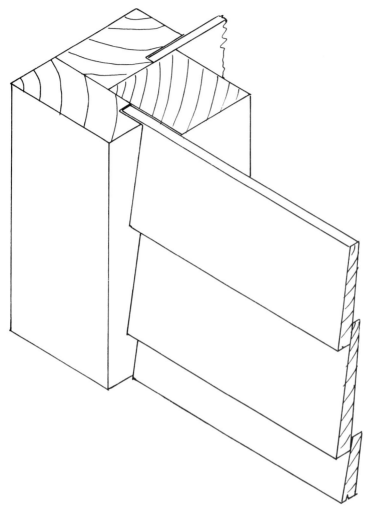

222 *Corner detail showing weatherproofing*

being taken with the edge details. A gutter should be fixed at the lower edge, either to the exposed ends of the joists, or, if preferred, by putting a fascia across the ends, and fitting the gutter brackets to this (fig. 220).

The door frames and window frames can be fixed, with attention being given to the way the feather-edge board and corner battens will be positioned. The boarding can be cut to length and carefully nailed, keeping the top edge horizontal throughout. The main vertical battens may be simply nailed in position, but it is far better if they are of a slightly thicker section, and scribed in after the boarding has been fixed (fig. 221). This will then provide a reasonable weather-tight joint as well as improving the appearance (fig. 222). The final finish will depend on the owner, but, as the shed will have to stand up to seasonal change, the traditional creosote may appeal, although care must be taken in its use as it is corrosive to both plants and skin. There are many proprietary brands of wood preservative available, some giving a coloured finish to the wood, and this may well be what is required.

134

FENCING

Types of fencing

Styles and construction methods of fencing are well-established, and hence most fencing can be classified under the following headings:

(a) cleft chestnut pale
(b) chain link
(c) post and rail
(d) palisade and close boarded
(e) ranch
(f) panel

The cleft or split chestnut pale is made by splitting short lengths of 35 to 50mm diameter (1½ to 2in) chestnut sapling into three, and then twisting galvanized wire around each pale. It may then be attached to straining wires or stapled to horizontal rails (fig. 223).

The interlaced chain-link fence may be galvanized or plastic coated, and again, is either fixed with tie wire to straining wires, or is stapled to horizontal rails. Timber, metal or concrete posts may be used (fig. 224). In both the above cases, if straining wires are used, then careful attention has to be given to the corner and end posts (fig. 225). This is generally done by providing a diagonal strut that acts as a prop to the post, or by placing a second post nearby with a top horizontal strut, and a wire straining tie from the top of the second post to the bottom of the first. On a long run, intermediate posts may also need some additional support.

223 Chestnut split paling

cleft (split) pole

twisted galvanised wire

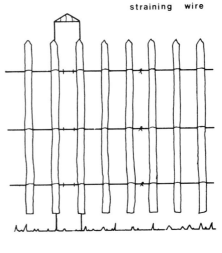

straining wire

fencing tied to straining wires

135

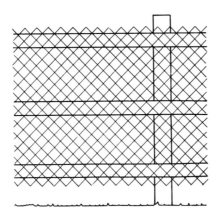

224 Interlaced chainlink

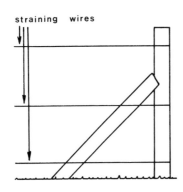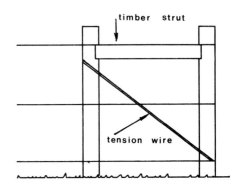

straining wires

timber strut

tension wire

225 Corner post details

226 Post and rail

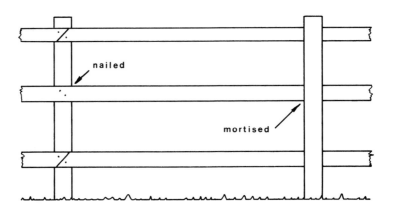

nailed

mortised

136

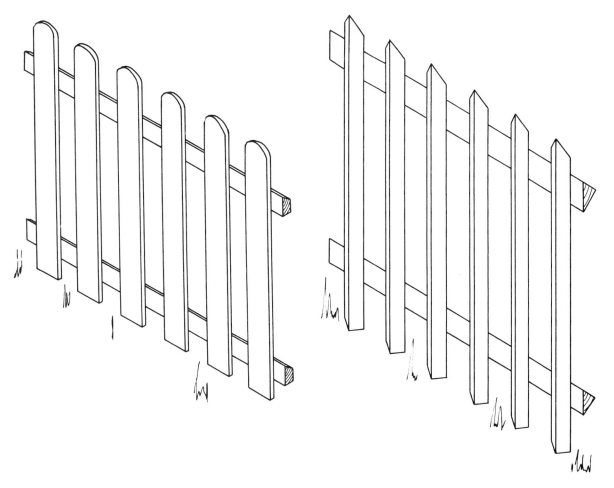

227 *Palings with rounded tops*

228 *Military paling*

Post and rail may have many variations, but the basic principle is for two or three horizontal rails to be fixed to vertical posts (fig. 226). The rails may be joined by scarf or butt joints, and they may be nailed to – or mortised into – the posts.

The palisade and close-boarded fences will provide a much greater degree of privacy, but, if they are to last for any reasonable period, they will require more attention to the design detailing as the end grain of each pale or board may be exposed to the elements.

Rails that are not properly protected will generally have the top surface bevelled, and, unless a weathered capping is provided, the tops of the palings or boards will be shaped to allow better run-off for rain and snow (fig. 227).

Military palings and arris rails are produced by ripping diagonally along a square sectioned timber, the palings being cut at 45° to produce a point (fig. 228). It is normal for the arris rails to be mortised into the timber support posts with this style of fence.

137

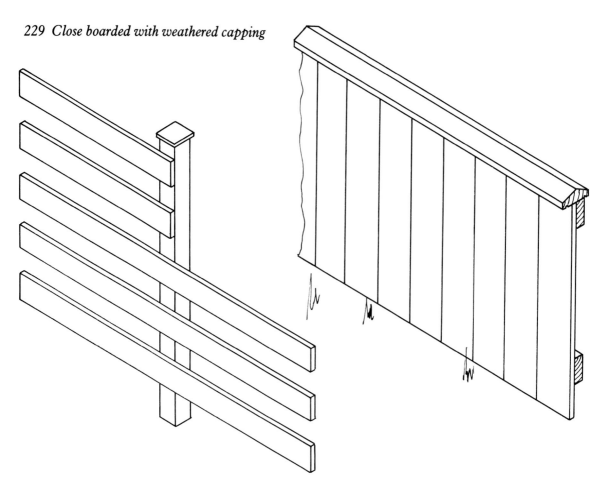

229 Close boarded with weathered capping

230 Ranch fencing

Close-boarded fences may be tongued and grooved, simply butted or lapped. Whichever method is chosen, a weathered capping should normally be added (fig. 229).

Ranch fencing consists of boards and gaps of equal widths, and they may be either vertically or horizontally fixed. The horizontal type consists simply of a series of rails, and is a more elaborate version of the post and rail. The vertical type requires both rails for support and a capping for the protection of the exposed ends of the vertical pales.

Panel fencing has become popular with the advent of modern housing estates and the small gardens associated with them (fig. 231). There are several types of panels available, either interwoven or with horizontal lapped boards, both types being prefabricated and simply requiring to be nailed to vertical support posts.

Great care must be taken to ensure that the distance between the posts accurately corresponds with the length of the panel, and that the posts are set perfectly upright.

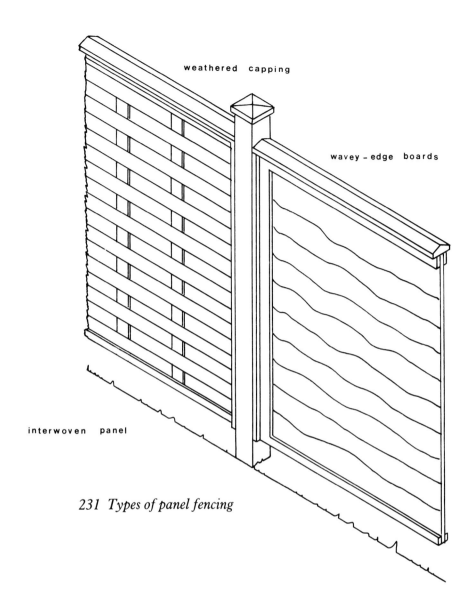

weathered capping

wavey - edge boards

interwoven panel

231 Types of panel fencing

Posts

There are various methods of sinking and anchoring posts, and the method chosen will depend on the type of fencing and the height, the material to be used for the post, and the depth of post in the ground. For a temporary fence, or where ground conditions are favourable, it may be possible simply to drive in a wooden stake and attach the fencing to it. Sloping ground may require a stepping method as, for the sake of appearance, the horizontal rails should be horizontal, and not sloped to correspond with the ground (fig. 232).

139

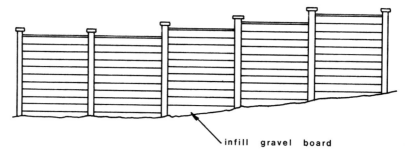

infill   gravel   board

232 Dealing with sloping ground

Holes may be dug with an earth auger or a spade, the auger providing a circular hole, the spade being used for square holes. A timber post hole may be backfilled with compacted earth, but if a more permanent post is required, then a mixture of concrete and broken brick will provide a suitable fill (fig. 233).

Unfortunately, over a period of time, timber posts in the ground will rot, and it may be preferred to set in a short concrete stump post to which the timber post may be attached using coach bolts (fig. 234). The tops of posts should be cut to provide a sloping surface, or the provided with a weathered capping to try to prevent decay by wet-rot (fig. 235).

Preservative treatment of all timber below ground is absolutely essential if a reasonable life expectancy is required for posts and fencing (fig. 236).

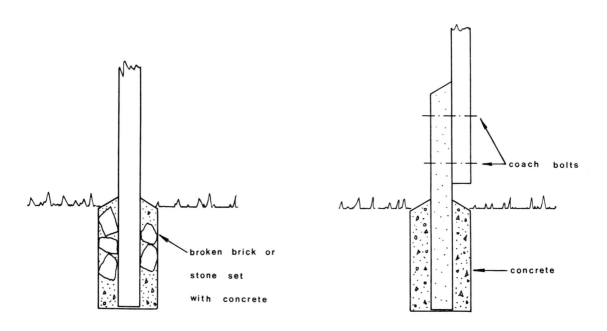

broken brick or

stone set

with concrete

coach   bolts

concrete

233 Typical fill to post hole   *234 Concrete stump post*

140

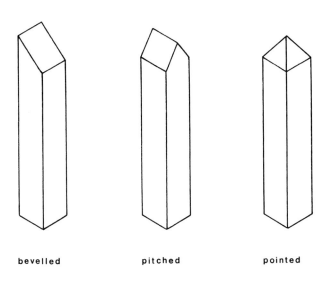

bevelled   pitched   pointed

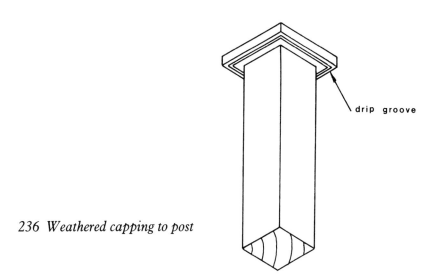

236 *Weathered capping to post*

drip groove

Materials

Although oak is the traditional material used for fencing, low grade soft wood is now generally used as there is a considerable difference in cost. However, this does mean that a great deal of time and care needs to be spent in treating the timber with preservative, and also allowing for regular maintenance throughout the life of the fence. Posts will normally be 75 × 75mm (3 × 3in) and generally have at least 600mm (2ft) sunk into the ground.

Arris rails will normally be 100mm (4in) on the face, and 75mm (3in) on the two sloping sides. Feather-edged boarding will be 100mm (4in) wide, and tapering from 20mm (¾in) to 6mm (¼in). When erected, the overlap should be about 25mm (1in).

141

INDEX